FOURIER SERIES AND OPTICAL TRANSFORM TECHNIQUES IN CONTEMPORARY OPTICS

FOURIER SERIES AND OPTICAL TRANSFORM TECHNIQUES IN CONTEMPORARY OPTICS

An Introduction

RAYMOND G. WILSON
Associate Professor of Physics
Illinois Wesleyan University
Bloomington, IL

With contributions by
Sean M. McCreary

A Wiley–Interscience Publication
JOHN WILEY & SONS, INC.

New York • Chichester • Brisbane • Toronto • Singapore

This text is printed on acid-free paper.

Copyright © 1995 by John Wiley & Sons, Inc.

All rights reserved. Published simultaneously in Canada.

Reproduction or translation of any part of this work beyond that permitted by Section 107 or 108 of the 1976 United States Copyright Act without the permission of the copyright owner is unlawful. Requests for permission or further information should be addressed to the Permissions Department, John Wiley & Sons, Inc., 605 Third Avenue, New York, NY 10158-0012.

Library of Congress Cataloging in Publication Data:

Wilson, Raymond G., 1932–
    Fourier series and optical transform techniques in contemporary optics: an introduction/Raymond G. Wilson: with contributions by Sean M. McCreary.
      p. cm.
    ISBN 0-471-30357-7
    1. Fourier transform optics. 2. Fourier analysis.
3. Diffraction. 4. Image processing. I. McCreary, Sean M.
II. Title.
QC454,F7W55 1995
535'.2'015152433—dc20   94-37922

Printed in the United States of America

10 9 8 7 6 5 4 3 2 1

*For Akiko,
And for the people of Hiroshima and Nagasaki
who have shed light for me on many other problems*

CONTENTS

Preface   xiii

1 Some of the How and Why of Fourier Analysis   1

   1.1 Periodic Functions in Time, 1
   1.2 Digression on Angular Terminology, 2
   1.3 Other Oscillations and Fourier Analysis, 4
   1.4 Periodic Functions in Space, 12
   1.5 Traveling Waves, 12
   1.6 Summary of Terminology, 14
   1.7 Who Was Count Jean Baptiste Joseph Fourier?
        "The Legacy of Fourier," by D. J. Lovell, 14
        Abbreviated References Appropriate to this Introductory
        Material, 20
        Topics for Further Consideration, 20

2 Fourier Series and Spectra in One-dimension for Functions of
   Finite Period   23

   2.1 Some Definitions of Fourier Series, 23
   2.2 Trigonometric Representation of Fourier Series, 26
   2.3 Exponential, or Complex Representation, of Finite Fourier Series, 37
   2.4 Trigonometric Fourier Series and Spectral Spike Rotation, 43
   2.5 Complex Fourier Series and Spectral Spike Rotation, 45
   2.6 A Shift in $f(x)$; Trigonometric Case, 48
   2.7 A Shift in $f(x)$; Complex Case, 51

2.8 Vertical Shifts, 52
Abbreviated References, 53
For the Reader to try, 53

3 **Fourier Series and Spectra for Functions of Infinite Period; One Dimensional** 54

3.1 Dealing with Period and with Pulse Width for Infinite Series, 54
3.2 Size of the Spectral Spikes, 59
3.3 Letting the Period, $2L$, Approach Infinite Extent, 65
3.4 Phase in Infinite Fourier Series, 74
   3.4.1 Phase Shifts When $2L = 2$, $p = +\frac{1}{2}$, 74
   3.4.2 Phase Shifts When $2L = 10$, $p = +\frac{1}{2}$, 77
   3.4.3 Phase Shifts When $2L = 10$, $p = +1$, 79
   3.4.4 Phase Shifts When $p = +\frac{1}{2}$, $2L$ Increasing, 80
   3.4.5 Phase Display for Fourier Series, 83

4 **Fourier Spectra for Non-periodic Functions; One-dimensional** 85

4.1 The Fourier Transform Pair for Functions of Infinite Period, 85
4.2 The Fourier Transform and Spectrum of a Square Pulse, 89
4.3 Phase in One-Dimensional Fourier Transforms, 92
4.4 The Real and Imaginary Parts of Fourier Spectra and Series, 96

5 **The Diffraction of Light and Fourier Transforms in Two Dimensions** 99

5.1 The Diffraction of Light, 99
5.2 The Fraunhofer Approximation and Two-Dimensional Fourier Transforms, 103
5.3 Explanation of Notation to be Used, 107
5.4 Diffraction Applications of the New Notation, 111
   5.4.1 Using the Angles of Diffraction, 113
   5.4.2 The Rectangular Aperture and Four Forms of the Fourier Transform, 116
5.5 Scale in the Fourier Transform Plane, 120
5.6 Making the Fourier Transform Accessible at Reasonable Distance, 121
5.7 The "Amplitude" of Optical Disturbances, 125
5.8 A Preliminary List of Some Properties of Fourier Transforms, 127
   5.8.1 Symmetry Properties of Fourier Transforms (One Dimensional), 127
   5.8.2 Linearity Property, 127
   5.8.3 Shifting Property, 127
   5.8.4 Scaling Property, 128
   5.8.5 Fourier Transforms and Complex Functions, 128

General References to the Diffraction of Light, 129
References to Diffraction in Radio Wave Propagation, 129

6 A Brief Summary of Linear Systems Theory Applied to Optical Imaging 130

6.1 Linear Systems Theory Applied to Optical Systems, 130
    6.1.1 A General Outline, 130
    6.1.2 Imaging, 132
        6.1.2.1 Coherent Imaging, 132
        6.1.2.2 Incoherent Imaging, 133
    6.1.3 Defining the Spread Functions, 135
    6.1.4 Definition of Optical Transfer Functions, 138
    6.1.5 Aperture Stop, Entrance and Exit Pupils, 141
6.2 Imaging with Incoherent Illumination, 142
    6.2.1 Image Formation with Incoherent Light via the *OTF*, Using the Spatial Frequency Concept, 143
    6.2.2 Image Formation with Incoherent Light via the *PSF*, Using the Impulse Response Concept, 145
6.3 Imaging with Coherent Illumination, 146
6.4 Transfer Functions by Convolution; Conceptual Basis, 148
6.5 Calculation Examples, 160
    6.5.1 Amplitude Impulse Response, 161
    6.5.2 Point Spread Function, 161
    6.5.3 Transfer Function for Imaging with Incoherent Light, 162
    6.5.4 Transfer Function for Imaging with Coherent Light, 165
References, 167
For Further Consideration, 167

7 Fourier Optical Transformations by Computer 169

7.1 Primitive Approaches, 169
7.2 Surface Plotting Hand-Calculated $Z(X, Y)$, 172
    7.2.1 The Square Aperture; Preparation for SURFER, 172
    7.2.2 The Rectangular Aperture; Preparation for SURFER, 175
    7.2.3 Circular Aperture; Preparation for SURFER, 177
    7.2.4 The Parallelogram Aperture; Preparation for SURFER, 180
7.3 Letting the Software Do It All; MATLAB Development, 186
    7.3.1 The Square Aperture, 186
    7.3.2 Other Real Apertures, 194
        7.3.2.1 Half Square, Half Transmitting, 194
        7.3.2.2 Circular, 196
        7.3.2.3 Square and Circle, a Composite Aperture, 199
        7.3.2.4 Triangle, 200
        7.3.2.5 Composite Apertures, Annular, 203
For Further Consideration, 208

x CONTENTS

8 Apodization and Super-Resolution, Phase from Shift, and Multiple Apertures — 209

8.1 Apodization and Super-Resolution, 211
 8.1.1 Apodization, 214
  8.1.1.1 Cosine Apodization, 214
  8.1.1.2 Linear Taper Apodization, 216
 8.1.2 Super-Resolution, 220
  8.1.2.1 Cosine Super-Resolution, 220
  8.1.2.2 Double Aperture Super Resolution, 224
8.2 Illustrating the Shift Theorem in Simple Diffraction: Phase in a Fourier Transform, 225
8.3 Multiple Apertures: Application of Shift and Linearity Principles, 229
8.4 Quality Apertures in the Laboratory, 241
 Questions and Challenges to Ponder, 242

9 Complex Apertures — 243

9.1 Phase Concepts, 243
 9.1.1 Half Aperture With π Phase Lag, 245
 9.1.2 Half Aperture With $\pi/2$ Phase Lag, 247
 9.1.3 Aperture With a Linear Phase Function, 250
 9.1.4 Aperture V-Linear Phase Function, 254
9.2 More Complex Aperture Functions; Aberrations, 260
 9.2.1 Coma, 262
 9.2.2 Astigmatism, 268
 9.2.3 Focus Error, 272
 9.2.4 Zernicke Polynomial Number 34, 273
9.3 A More Complete Summary of Basic Fourier Theorems, 278
 9.3(a) Linearity Theorem, 278
 9.3(b) Shift Theorem, 279
 9.3(c) Modulation Theorem, 279
 9.3(d) Scaling Theorem, 279
 9.3(e) Separation Theorem, 279
 9.3(f) Radial Functions, 279
 References for Wave Aberrations, 280
 For Further Consideration, 280

10 Operations in the Fourier Transform Plane — 281

10.1 Step Into the Fourier Transform Plane, 283
10.2 Edge Detection by Spatial Filtering, 284
10.3 Spatial Filtering for Detection of Aperture Phase, 286
10.4 Results From the Spatial Filtering Laboratory, 290
10.5 An Inverse Problem, 291
10.6 Some Personal Conclusions, 302

11 Other Interesting and Related Topics   **304**

11.1 Laser Beacon Adaptive Optics, 304
11.2 Fresnel Diffraction Via Fourier Transform, 307
References to Fresnel Diffraction by Fourier Transformation, 310

References   **311**

College Level Optics Textbooks which Contain Material on Fourier Methods, 316

A Selected Bibliography   **317**

Index   **323**

PREFACE

Although the professional researcher who daily uses Fourier transform techniques and professors of optics or image processing will find in this book some graphic portrayals which they have used, discussed, taught, but probably never seen before, this book is really directed toward the novice at Fourier methods. As such, it could be used as the text for a *first* course in Fourier techniques. The reader will find it rich in illustration, examples, and explanations, in definitions and properties, in 2- and 3-space, but ultralight in lemma, theorem, proof, and corollary. Thus do I suggest class time activities to the teacher.

Because of its 3-space portrayals, this book could become *a supplementary reference*, for teacher or student, to *any course at the undergraduate level dealing with Fourier series and transforms, and with the convolution theorem*. I have placed great emphasis on graphical portrayal of the series and transforms. You will find the abundant illustrations, and how to do them, rather unique.

Another audience for this book would be the science or math student who has completed a good course in General Physics and a semester or two of beginning Calculus; someone in school now, or someone returning to this level for new understanding. I've used some quite basic (in 1995) computer software for the graphics, so you should be prepared for the "Wow, look what I can do!" experience.

The career field of *opto-electronics* seems to usually not suffer too much from employment problems. Fourier techniques are fundamental to much of the day-to-day work in this field and also on its frontiers. And it's fun to do!

In mathematical level the writing is roughly similar to the earlier excellent monographs: E. G. Steward's *Fourier Optics: An Introduction*, 1983; C. Taylor's *Diffraction* (Student Monographs in Physics), 1987 and his *Images—A Unified View of Diffraction and Image Formation with all Kinds of Radiation*, 1978; and D. C. Champeney's *Fourier Transforms in Physics* (Student Monographs in Physics), 1985. However, the visualizations in the present book have never, or at most very rarely, been seen in a book written for understanding by undergraduates.

Some systematic collections of Fourier graphics in 3-space have been published, notably three articles in *Applied Optics and Engineering*, but all three leave out some aspect which would make for completeness. Those authors did not intend the completeness I sought nor were their articles intended as course text. None of the three give any hint as to "how it was done," except, obviously, by computer. "How do you do that?" is important to me.

For the user one beauty of the Fourier series and transform approach is its reliance on sinusoidal functions and their equivalent exponential expressions. These are well-known functions. You'll find a great deal done here without becoming more sophisticated than that. So we're ahead of the game already at the beginning.

I've tried to write a friendly, somewhat chatty book, trying to convey some of my own enthusiasm and pleasure, as when certain plots worked out well or revealed something unsuspected. I wanted the book to be readily master-able, like those of Taylor, Steward, and Champeney, and I wanted the graphics to be eye-catching and in a sense, for you, "How did he do that?" And I tell how. My intention was to stretch your mind into a new dimension and allow you visualizations that the eye cannot see. I want you also to realize, yes—here is a computer technique used in diffraction phenomena, but these ideas could be used in my psych. lab, or my math research, or my materials science project. You'll find I take considerable liberties in Chapter 1; it is folksy; as is material in a later chapter, with references to screwed-up or miss-filtered communications signals and to Hirohito; signal analysis takes place in more domains than the science and technology labs.

The software Sean McCreary and I used, that of Golden Software and of the MATHWORKS is continually updated and improved to meet the challenges of the latest decade. These computer graphics capabilities did not develop overnight.

In the first four chapters are introduced the mathematical terminology for wave phenomena with emphasis on spatial frequency terminology. The concluding section of Chapter 1 is a brief biography of Fourier by the late D. J. Lovell. If the movie is made of Fourier's life Harrison Ford will probably get the role. Chapters 2 through 4 deal with Fourier series and spectra of one-dimensional functions, periodic and nonperiodic. Both trigonometric and exponential forms are used and thus the Euler relationships are introduced and used. Chapter 3 focuses more on physical concepts rather than any new mathematical details. The Fourier transform makes its first appearance in

Chapter 4 to deal with nonperiodic functions. Much of the graphics has rarely been seen before.

The 3-space spiral plots of Chapters 3 and 4 are about the only way to incorporate into a *single* 3-space diagram the concepts of phase AND amplitude for Fourier series and transforms of one dimensional signals. As I point out it is impossible to do this, phase AND amplitude, in a 3-space diagram for analysis of 2-space signals; it leads to separate phase diagrams, onward from Chapter 7. The spirals of Chapters 3 and 4 are similar to much earlier hand-drawings employed by Duffieux, Bracewell, Gaskill, and no doubt, others. It is a lovely technique, one you perhaps have not seen before, difficult to do well by hand, not all that easy by computer, especially when the amplitude of the spiral must change in a prescribed manner. I wanted the concept of phase to be not lightly treated, and this was how I chose to "picture" it.

Chapter 5 develops the relationship between Fraunhofer diffraction by two-dimensional apertures (functions) and the Fourier transform in 3-space. Here I use care in explaining the notation to be used, the abstraction now having jumped one more step, from 2-space. Examples are presented, in 3-space, and a preliminary list of Fourier transform properties is presented. A more complete summary of transform properties in 3-space is in Chapter 9. Since this book was not intended to be thorough in all properties of Fourier transforms, as in the advanced books of Gaskill, Bracewell, Walker, etc., these properties are only summarized, but summarized well.

The next large step in conceptual/mathematical extension is the employment of the convolution theorem in Chapter 6. This has been the most difficult chapter to write—for this level book. My attempt is through "convolution-theorem flow diagrams," using the concepts primarily developed in Chapter 5 and words, to explain the use of the convolution theorem as it applies to Fraunhofer diffraction and to imaging. You will probably not find mathematical convolution in 3-space easy to do, and only a few simple convolutions of meaning will be mathematically do-able. At this level it may be best to leave convolution for the software to do, assuming you understand the geometrical concepts (for real apertures). The "delta function" was purposely not introduced nor the array theorem as such, though diffraction by arrays of identical apertures is demonstrated. For the imaging concepts of Chapter 6 I believe the use and meaning of convolution has been "explained" well.

Chapter 7 describes how the 3-space transform graphics were done by computer. One approach is simply a hidden line (or unhidden line) surface plotter, for Fourier transforms which have been calculated by hand as $z = f(x, y)$. For this I use SURFER ®. Great satisfaction can be obtained here as one sees coming to fruition the result of what might have been a challenging calculation. Amplitude and phase can be displayed, something not ordinarily visually possible in the lab. Very recently instruments such as the Melles-Griot *WaveAlyzer*™, which instrument also relies on computer processing of lab input data, can do a 3-space display of wavefront phase. The second computer approach McCreary and I used is through software such as MATLAB® which can do Fourier transforms of two-dimensional functions and display the result in 3-space. MATLAB is capable of all the operations present in the convolution theorem (and in signal processing), including inverse transformations. It operates upon one- and two-dimensional functions which must be defined by the user. If you can define it so MATLAB can understand it, you can let your imagination run wild in the realm of real and complex functions. Great fun! Our attention is focused on *aperture* or *pupil* functions.

The interesting possibilities of apodizing and super-resolving aperture functions are examined in Chapter 8. Phase in the transform is once again considered as are multiple aperture functions.

Complex aperture functions are the focus in Chapter 9 were McCreary and I start easy but finish with coma, astigmatism and Zernicke Polynomial Number 34. Here also is the summary, for use in 3-space, of Fourier transform properties.

Chapter 10 steps into the transform plane to manipulate the spatial frequency spectrum there and hence "enhance" images formed through such "spatial filtering." Since MATLAB can do inverse transforms an "inverse problem" is considered. A special flavour is added by the inclusion of some very fine lab work, from the United Kingdom, of G. Harburn, C. A. Taylor, and T. R. Welbery. It is, of course, these kinds of operations, not necessarily with visible light, which have revealed the structures of DNA and other biologically important molecules, and other kinds of pico-structures which cannot actually be seen in order to be measured.

Now that military secrecy has been lifted much of the work in laser beacon adaptive optics is now reported in the literature. That topic fits nicely into Chapter 11 and I'm amazed at the results. Also here, is a brief description of Fresnel diffraction by Fourier transform techniques, avoiding the use of the Cornu spiral, which use always had severe limitations to quite simple aperture functions. It is a certainty that more will be done with this, especially now that computers have such amazing capabilities. Even so, computers can only answer questions which we have the wits to formulate for them.

Sources of high-quality inexpensive diffracting apertures are mentioned in Sections 10.6 and 8.4. Software sources including SURFER and MATLAB are referred to at the end of Chapter 7.

This work had its origin via NSF Grant SED-8021473. Jeffrey L. Thompson participated in the early stages. David Botkin also made early contributions. Judy Switzer typed most of the early manuscripts, but I typed all of the final ones using Chiwriter 3.0. There are a considerable number of mathematical calculations and computer operations represented here. Sean McCreary worked seemingly tirelessly on many of these but R. Wilson takes responsibility for all the errors.

I would greatly appreciate comments about this book from its users and from its potential users. What did you find missing, what should be left out? Where do I need to rewrite? What have you tried that worked out well and that might be of benefit to others? What are some good problems to add? Should the Gibb's phenomena be added? We might be able to include additional insights in a second edition. You can write to me at the Physics Department, Illinois Wesleyan University, Bloomington, IL 61702. I will gladly respond to every communication.

The editors at John Wiley, Gregory Franklin, John Falcone, Rick Mumma, Kimi Sugeno and their staff have my gratitude for greatly improving the presentation of this material.

There are a number of people in Japan who, through their help, friendship, and trust in me, have made possible the expansion of my own personal horizon of understanding: Masuyuki Imaishi, Naomi Shohno, Yuki Setoyama, Hideaki Nagai, Miyoko Matsubara, and really, many others. Domo arigato gozaimashita.

I am grateful for the affection, never ending encouragement and support, patience, and understanding from Carolyn, Laura, Tim, and Dave. I've often wondered if it was true, as other writers have claimed, that we neglect our families as these writing projects consume time and us... It is... As a physicist I can't claim to believe in fate; but how else can I explain Akiko's presence in my life, from halfway around the world? From the beginning, she has at one, been the perfect complement and supplement to my life, enabling me in more ways than I can number. I hope she feels the same. The gift of Aya and Taiyo only she could give. Thanks, you three for getting me off to work each day; three each for each of you. Now the vacation?

Raymond G. Wilson

Normal, Illinois
September, 1994

1

SOME OF THE HOW AND WHY OF FOURIER ANALYSIS

This chapter is a very brief introduction to some of the terminology of periodic functions. It is also an elementary overview to stress the fact that Fourier analysis can be applied in many different areas of the natural world, and the world created by man. To keep it simple here we will refer mainly to one-dimensional situations, but the thrust of this book is into the realm of Fourier analysis of two-dimensional functions, where we will also find wide application in the natural and man-created worlds.

1.1 PERIODIC FUNCTIONS IN TIME

When simple objects vibrate or oscillate some aspect of the motion can usually be described by a simple sine function. Examples of such motion are a mass oscillating on a spring, a vibrating guitar string. It is inherent in the physical nature of these oscillations that, to a very good approximation, the motion can be described by sinusoidal functions. The physical laws which apply to oscillating objects, e.g., Newton's laws, lead to mathematical expressions which represent the actual motion. One such representation might be

$$x(t) = A \sin(t)$$

Consider then the sinusoidal oscillation in time pictured in Figure 1.1. Mathematically we describe it as:

$$f(t) = A \sin(\omega t) \quad \text{or} \quad A \sin(2\pi f t) \quad \text{or} \quad A \sin(2\pi t/T)$$

2 SOME OF THE HOW AND WHY OF FOURIER ANALYSIS

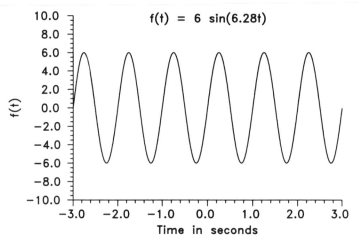

Figure 1.1 Graphic portrayal of an oscillation in time.

T is the time it takes to complete one period of the oscillation, here 1.0 second; T is called simply "the period of the oscillation," it is the period in time, the temporal period. The frequency of the oscillation, f, is the number of oscillations per unit time, here 1.0 per second; f is the temporal frequency. Do not confuse it with the $f(t)$ where f = "function."

The maximum amplitude of the oscillation is A, here equal to 6. $f(t)$ represents the instantaneous displacement of the oscillation from the zero position. A and $f(t)$ will both have the same dimensions (e.g., pressure, or length or power) and therefore would require the same units of measure (e.g., Newtons/m^2 or meters or Watts). People, sometimes using rather loose syntax, call this curve a sine *wave*, but it does not move in space; though your eyes may when viewing the function. The curve represents only an oscillation in time. A disturbance (function), sinusoidal or otherwise, *moving through space* is a wave. Last to be defined is ω.

1.2 DIGRESSION ON ANGULAR TERMINOLOGY

The quantity ω is called the angular *frequency* of the oscillation, although you can see that there are no angles involved here. The terminology arose originally from circular motion, for example, the axial rotation of a wheel, where ω was the angular *velocity* of the wheel usually expressed in units of radians/second. (If these units are unfamiliar to the reader, angular velocity can also be expressed in units of revolution/minute, such as $33\frac{1}{3}$ rpm; or units of angular degrees/second.) Angular *speed* is often taken to mean the same thing as angular *velocity*. Consider such a wheel, then, rotating with an angular velocity ω; the vertical motion alone or the horizontal motion alone of any *point* on the wheel can be described by a simple sine function of frequency $f = \omega/2\pi$. For example,

$$x(t) = A \sin(2\pi f t) = A \sin[2\pi(\omega/2\pi)t] = A \sin(\omega t)$$

would be the motion in x described by a trigonometric function of ω and t. The expression $B \sin(\omega t)$ could be the motion in y, a similar function of ω (radians/second) and t. The x and y motions are described using an angular quantity, though the actual motion is non-angular. The quantity ω, having dimensions of angle per time, is thus properly termed the temporal *angular* frequency. It must be clearly distinguished from the temporal frequency f.

Cosine functions can also be used, depending on when both x and t are equal to zero, e.g., $y(t) = C \cos(\omega t)$. In any case, when a wheel rotates at an angular *velocity* (or speed) ω, any *point* on the wheel performs sinusoidal motion in both horizontal and vertical directions (as well as in other directions). The x (or y) displacement of the point is in a straight line, back and forth, but straight (linear). Dimensionally we see that:

$$\omega \left[\frac{\text{radians}}{\text{seconds}} \right] = \frac{2\pi}{T} \left[\frac{\text{radians/cycle}}{\text{seconds/cycle}} \right] = 2\pi \left[\frac{\text{radians}}{\text{cycle}} \right] \cdot f \left[\frac{\text{cycles}}{\text{second}} \right]$$

$$\omega = 2\pi/T = 2\pi f$$

Thus it is hoped that the reader understands how angular measure crept into the description of the linear motion of simple oscillations.

In the above discussion, other than the amplitude and displacement, x, A, or y, we have referred to only one other quantity which has a proper physical dimension, time. Angular measure, such as radians, are ratios of lengths and hence dimensionless;

$$\theta[\text{radians}] = s[\text{arc length}]/r[\text{radius}]$$

"Angular degrees" and "cycle" and "revolution" are also inherently dimensionless quantities. But if we did not use specific units of measure for angular quantities there would be much confusion as to whether one was speaking of radians, degrees, revolutions, or what.

Therefore it is highly desirable to refer to angular frequency, ω, in specific units such as [radians/second] or [degrees/second] rather than just: per second, 1/second, or second^{-1}. Likewise the temporal frequency f which now has official units of Hertz could also be referred to with units of [cycles/second] or [revolutions/second]. Do you see that angular velocity and angular frequency can be properly expressed with units of second^{-1}? But unless the context is clear, confusion could result. We will, therefore, consistently designate

4 SOME OF THE HOW AND WHY OF FOURIER ANALYSIS

units as completely as possible, e.g., [radians/second], or [degrees/second], or [revolutions/second].

1.3 OTHER OSCILLATIONS AND FOURIER ANALYSIS

In addition to simple sinusoidal functions there are other kinds of periodic oscillations in time, e.g., those which might be called a "square wave," a "triangular wave," or a "sawtooth wave"; most readers are probably familiar with their appearance (Figure 1.2). (From this point on we will occasionally use the expression "wave" even though the oscillation is a function of a single variable, x or t. The context should make the meaning clear.) How would you write a mathematically analytic expression for, say, a function which looks like a square wave?... Fourier analysis (or spectral, or harmonic analysis) indicates that with few exceptions *any* periodic function can be fairly well mathematically represented by the sum of a series of sinusoidal terms. There are special cases of periodic functions (or waves) which can be represented by the sum of a finite number of terms, a finite series of terms. But in many cases, for perfect representation, in theory, an infinite number of terms would be required, an infinite series. But in practice one truncates, cuts off the series and thus uses a finite and sufficient number of terms as an approximation. Thus a square function can be analytically represented approximately by a mathematical series with a finite number of terms. If one sums the series one produces (approximately) the square function. These are called Fourier series.

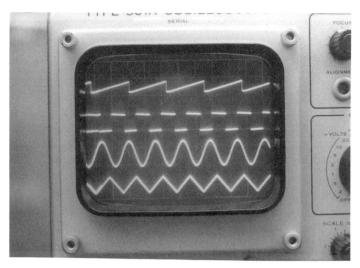

Figure 1.2 Some electrical periodic signals displayed on an oscilloscope: a "sawtooth" wave, an (approximately) square wave, a sine wave, a triangular wave.

OTHER OSCILLATIONS AND FOURIER ANALYSIS 5

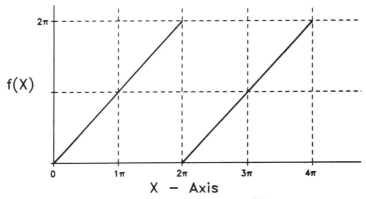

Figure 1.3 A mathematical "sawtooth" pattern.

Here is an example taken from a mathematical handbook. It says that the wave shown in Figure 1.3 can be represented by this expression,

$$f(x) = \pi - 2\left(\frac{\sin x}{1} + \frac{\sin 2x}{2} + \frac{\sin 3x}{3} + \cdots\right)$$

We'll show how to find such series shortly. If you have access to a computer that can plot $f(x)$ as a function of x you might want to try letting it draw this wave. You can see how to add more terms; the more terms used, the better the approximation. Figure 1.4 is our attempt, out to the $\sin(6x)$ term.

Since the terms of the series are simple analytic functions, sinusoidal functions, mathematically operating on the series of terms is equivalent to

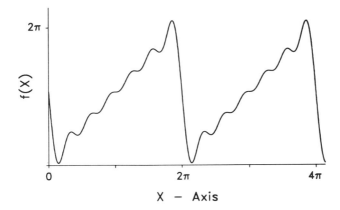

Figure 1.4 Adding together the first seven terms of the Fourier series representation for the sawtooth pattern of Figure 1.3.

operating on the square function (or wave) itself and opens the possibility of operating on selected terms of the series which corresponds to filtering the actual wave. We will illustrate this shortly.

The Fourier series approach is not the only way to mathematically represent a periodic wave; representation can also be done with other series, e.g., a series of polynomials named after the mathematician, Legendre. The elegant simplicity inherent in the Fourier approach is: its use of only simple sinusoidal terms, sines, and/or cosines, familiar to high school physics students; and the relative ease with which one can specify each term in the series.

Consider a complex sound wave at a microphone, air pressure changing in time. Such a wave can be easily displayed as an electrical signal on an oscilloscope. That wave can be Fourier analyzed into its harmonic components, its series terms, mathematically, or instrumentally. A computer can be programmed to do the analysis but there are also electronic instruments (Figure 1.5) specifically designed for this job. In earlier days electrical and mechanical devices did such analysis. Each term, in the Fourier series which represents the wave, corresponds to what we call one of the harmonics, or overtones, of the original sound wave. (The musician will recognize this terminology.) A device which takes a periodic wave and separates out the individual terms (harmonics, or overtones) is doing Fourier analysis (or spectral or harmonic analysis). We use the word "spectral" here because each individual harmonic term has its own frequency of oscillation and as such each term represents a particular wavelength, or in light wave sense, color.

Charles A. Culver, in his book *Musical Acoustics* (1956), showed the results of this work analyzing the harmonic content of sound from musical instruments. In Figure 1.6 are the waveforms from the four open strings of a violin. Figure 1.7 illustrates the relative harmonic content of the signals seen in Figure 1.6.

Since sound waves travel in time and space they follow the simple distance equals velocity multiplied by time relationship:

$$d = vt$$

In a time T, the period of one oscillation, the wave travels a distance of one wavelength, λ, at velocity v, so:

$$\lambda = vT = v/f \qquad T[\text{seconds/cycle}] = 1/f[\text{cycles/second}]$$

Sound waves of long wavelength (low frequency) may have wavelengths of about 10 meters. High frequency sound waves have lengths of a few millimeters. Visible light is composed of waves of even smaller wavelength.

White light is composed of light waves of lengths ranging from 0.0000004 meters to 0.0000007 meters, from what we have come to call a violet color to

OTHER OSCILLATIONS AND FOURIER ANALYSIS 7

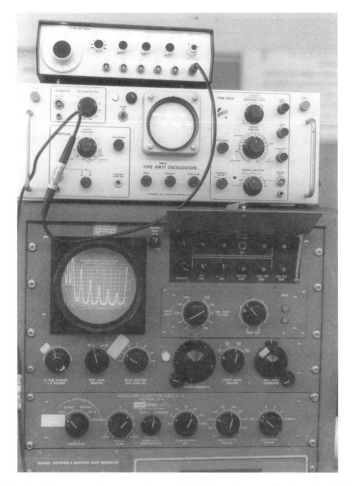

Figure 1.5 An older sound wave analyzer displaying the amplitude of each of the first few terms of an electronically produced square wave.

what we sense as a red color. A spectroscope takes all the waves in white light and spreads them out (disperses them) into a spectrum, the long waves, low frequency red light, at one end of the spectrum and the short waves, high frequency violet light, at the other. Hence a spectroscope is doing Fourier (spectral) analysis (Figure 1.8). You see why the expression "spectral" is appropriate. A spectrum is visually observable when dealing with light waves; musicians sometimes use the term "spectrum" or "color" in a somewhat different sense when referring to music.

All readers who have functioning sensory systems possess several types of Fourier analyzers. In the cochlea of the ear along the length of the basilar

8   SOME OF THE HOW AND WHY OF FOURIER ANALYSIS

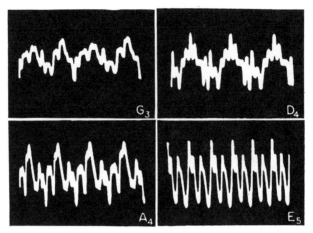

Figure 1.6 Waveforms from the four open strings of a violin. (C. A. Culver, *Musical Acoustics*, © 1956, reproduced with permission of the publisher, McGrawHill.)

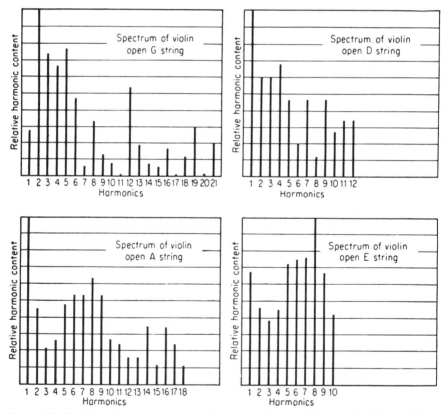

Figure 1.7 Relative harmonic content of the waveforms of Figure 1.6. (C. A. Culver, *Musical Acoustics*, © 1956, reproduced with permission of the publisher, McGraw-Hill.)

OTHER OSCILLATIONS AND FOURIER ANALYSIS 9

Figure 1.8 A rather large spectroscope for the spectral analysis of light.

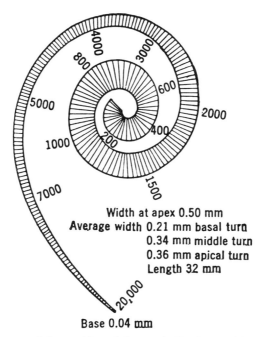

Figure 1.9 Diagram of the cochlea of the ear indicating positions along the cochlea most sensitive to different frequencies of sound. (Otto Stuhlman, Jr., *An Introduction to Biophysics*, copyright © 1943, reprinted by permission of the publisher, John Wiley and Sons, Inc. Rights controlled by Mrs. William T. Couch.)

Figure 1.10 One can sense by touch the spatial frequency of the grooves on a phonograph record.

membrane certain positions are stimulated only by certain wavelengths and hence the cochlea (Figure 1.9) spatially separates out individual harmonics of a complex sound wave. The brain "resynthesizes" them, the entire system, ear, nerves, and brain, doing it adequately but never perfectly. No one has a hearing system with a "flat" response throughout the entire sound spectrum. Of course, we are referring here only to sound within the normal hearing range.

Since the eye and brain can differentiate between red and blue light, and other colors, the eye and brain act as a spectral analyzer; we say color-blind people have faulty spectral analyzers.

Humans (and some animals, if trained properly) have the ability to analyze spatially periodic functions. The grooves on a phonograph record (Figure 1.10) are more closely spaced (higher spatial frequency, grooves/millimeter) than are the teeth of a comb (Figure 1.11), or the pickets in a fence (Figure 1.12). The eye/brain system as well as the tactile/brain system can sense and separate out these differences in spatial frequency. We can associate with these frequencies, spatial periods, i.e., wavelengths. Frequencies might be expressed as: [grooves/millimeter], [teeth/inch], [pickets/foot], and these are properly termed spatial frequencies. The spatial periods, i.e., wavelengths, might be expressed as: [millimeters/groove], [inches/tooth], [feet/picket].

OTHER OSCILLATIONS AND FOURIER ANALYSIS 11

Figure 1.11 The spatial frequency of a comb's teeth is much lower than that of a record's grooves.

Figure 1.12 The young woman is about to measure the spatial period of the fence pickets.

1.4 PERIODIC FUNCTIONS IN SPACE

We have referred to sinusoidal *functions of time*:

$$f(t) = A \sin(\omega t) = A \sin(2\pi f t) \quad \text{or} \quad A \sin(2\pi t/T)$$

and we have briefly discussed the analysis of other functions periodic in time. But now we see that we could also have functions periodic in space, e.g., the picket fence. And for now, if we just consider the one-dimensional square-wave nature of the pickets, i.e., picket, open, picket, open, picket, etc., we can analogously describe Fourier analysis of this spatial function. Here each term would be a sinusoid, not in time, but in space, such as:

$$f(x) = A \sin(kx) = A \sin(2\pi s x) \quad \text{or} \quad A \sin(2\pi x/\lambda)$$

The notation is exactly analogous to the terminology used for $f(t)$: λ is the spatial wavelength analogous to the temporal period, T; λ is the spatial period. Units of λ would be [meters/cycle] or [meters/wave] or simply [meters]. s is the spatial frequency analogous to the temporal frequency, f; units of s would be [cycles/meter] or [waves/meter] or simply [meter^{-1}]. The symbols r and s are recommended for spatial frequency; in these early chapters we use s for obvious reasons. k is the spatial angular frequency analogous to the temporal angular frequency, ω, although again no angles are involved and "angular" is a misnomer due to historical development, or due to the fact that angular functions happen to be a good describer of these spatial phenomena. The units of k would be [radians/meter] or simply [meter^{-1}]. Hence one sees again why one must be careful to specify whether one means spatial *angular* frequency k[radians/meter], or spatial frequency s[waves/meter] or [cycles/meter]; both could legitimately have units of [meter^{-1}] since radians, waves, and cycles are dimensionless quantities.

There is no standard symbol for spatial angular frequency, though k is commonly used in America when working with one-dimensional signals. Often, the careful distinction between s and k is not made, though it should be. Be careful not to confuse k, the spatial angular frequency, with that same symbol and its many other meanings: Boltzmann's constant, Coulomb's constant, a Hooke's law constant, etc.

1.5 TRAVELING WAVES

A real oscillation in space must also be an oscillation in time. Light can be considered such an oscillation; we describe it in more detail in Chapter 5. But an oscillation in the y-direction, e.g., a mass on a spring moving in time, provokes nothing apparent in, say, the x- or z-directions. But let the mass on the spring be oscillating as $y(t)$ in and out of a water surface and traveling waves will propagate out in the x–z plane water surface.

The velocity of the waves, v, is dictated by the physical properties of this particular fluid, e.g., the "flexibility" of the fluid to the mass forcing it to move. The surface wave disturbance travels one wavelength, λ, in time T, the wave period, at velocity v; hence from $d = vt$,

$$\lambda = vT$$

The driver of this oscillation is the mass on the spring of period T.

If you concentrate on *one point* at the water surface, the disturbance, its "average motion," will seem to be up and down in time like $y(t) = A \sin(2\pi t/T)$. This concentration is sometimes difficult; your eyes like to follow the wave and in doing so you "put yourself" in the moving reference frame of the wave. If you photograph the surface at a single instant of time the photo will reveal the space nature of the wave such as

$$y(x) = A \sin(2\pi x/\lambda)$$

The actual traveling wave disturbance is a combination of both the time and space aspects,

$$y(x, t) = A \sin(2\pi x/\lambda \mp 2\pi t/T) = A \sin 2\pi(x/\lambda \mp t/T)$$

In this form it is easy to see that if x equals zero or some other constant position, $y(x, t)$ will yield just the time oscillation at that position. While if t equals zero or some other constant time, $y(x, t)$ will yield just the x-space oscillation at that time. The minus sign of $\mp$ indicates motion in the $+x$ direction; the plus sign indicates motion in the $-x$ direction. Utilizing the water wave properties, from $\lambda = vT$, and the other relationships we have discussed, this expression can be written several ways.

$$y(x, t) = A \sin 2\pi(sx \mp ft)$$
$$= A \sin(kx \mp \omega t)$$
$$= A \sin k(x \mp vt)$$
$$= A \sin 2\pi f(x/v \mp t)$$

We can put this into the context of light waves traveling in free space where the velocity v is about 3×10^8 meters per second, and is represented by the symbol c. Hence

$$y(x, t) = A \sin k(x \mp ct)$$
$$= A \sin 2\pi f(x/c \mp t)$$
$$= A \sin(kx \mp 2\pi ct/\lambda)$$

Often an arbitrary direction, r, is used in place of x. For a light wave we have yet to describe the nature of $y(x, t)$ or A, surely not a displacement.

1.6 SUMMARY OF TERMINOLOGY

We summarize:

Functions of Time	Functions of Space
Temporal angular frequency, ω[radians/second] $= 2\pi f$	Spatial angular frequency, k[radians/meter] $= 2\pi s$
Temporal frequency, f[Hertz] or [cycles or waves/second] $= 1/T$	Spatial frequency, s[cycles or waves/meter] (no standard unit like Hertz) $= 1/\lambda$
Temporal period, T[seconds/cycle] $= 1/f$	Spatial period or wavelength, λ[meters/cycle] $= 1/s$
$\omega = 2\pi f = 2\pi/T$	$k = 2\pi s = 2\pi/\lambda$

Of course it is proper to use other units of time and length if done consistently, e.g., years and parsecs, fortnights and furlongs. The above are generally understandable MKS units.

The above will not be adequate when we move into 3-space and we will clearly make definitions later, but we can introduce the terminology here to simply show the analogy. We will need two sets corresponding to each of the x and y directions; z we reserve for the magnitude of the "optical disturbance" itself. The spatial frequencies will be represented by ξ and η.

In the x domain: $u = 2\pi\xi$. (ξ = xi, pronounced tzee or tsee)
In the y domain: $v = 2\pi\eta$. (η = eta, pronounced ate-ah)

The "periods" will be defined by geometry yet to come.

1.7 WHO WAS COUNT JEAN BAPTISTE JOSEPH FOURIER?

There are at least two book references in English to the life and times of Joseph Fourier, Herivel's *Joseph Fourier—The Man and the Physicist*, and *Joseph Fourier 1768–1830* by Grattan-Guinness. We are grateful to the Society of Photo-Optical Instrumentation Engineers (SPIE), the copyright holders of the following article, composed by the late D. J. Lovell, for permission to include it

Figure 1.13 Portrait of Joseph Fourier. (Courtesy of Smithsonian Institution Libraries, from negative number 91-417.)

in this book. The text is from the recent SPIE publication, edited by Frederick Su, *Technology of Our Times: People and Innovation in Optics and Optoelectronics*, SPIE Optical Engineering Press, Bellingham, WA, 1990.

If you read between the lines you'll find here the subject for a good historical novel. For his times, perhaps we could call Fourier "the intellectual's Indiana Jones."

THE LEGACY OF FOURIER©

By D. J. Lovell

D. J. Lovell, whose long career in optics included periods at MIT, Norwich University, the Naval Research Laboratory, and Los Alamos Scientific Laboratory, was the author of Optical Anecdotes, published by SPIE (Society of Photo-Optical Instrumentation Engineers) in 1981. The popular collection of profiles of noted "optikers" has been reprinted several times. This article appeared as the 20th anecdote in the book.

In his eulogy of Joseph Fourier to the Paris Academy of Sciences, Francois Arago concluded, "My object will have been completely attained if... each of you has learned that the progress of general physics, of terrestrial physics, and of geology will daily multiply the fertile applications of *Théorie de la Chaleur*, and that this work will transmit the name of Fourier down to the remotest posterity." Although Arago thus predicted a legacy of the use of Fourier's mathematical treatment used to describe the conduction of heat, in 1833 he could hardly have envisaged those benefits to be derived in optics. Today, Fourier optics and Fourier transform spectroscopy are widely practiced by scientists with little knowledge of the life of Joseph Fourier.

It is of interest to review Fourier's life and to reflect on the manner of man who brought to us a keener understanding of optical phenomena. Much of my information about his life is derived from the aforementioned eulogy, which was written by a friend and admirer. It was also written after a turbulent period in French history when the passion of patriotism gave rest to overindulgent attitudes toward behavior. I cite this to indicate that I have tried to separate fact from fancy, and to warn you that my attempts to present a clear account of Fourier's life may, at times, be colored by the biases introduced into the documents available to me.

Fourier was born at Auxerre, about 160 kilometers southeast of Paris, on March 21, 1768; he was orphaned at the age of eight. A neighbor lady, who recognized his courteous manners and his precocious natural abilities, recommended him to the Bishop of Auxerre, and Fourier was admitted to the military school run by the local Benedictines. His precocity was soon evident, as Fourier anonymously authored many of the sermons delivered by high dignitaries of churches in Paris.

Although evidently a gifted child, Fourier was also petulant, noisy, and vivacious. However, during his fourteenth year he became interested in mathematics and settled down (or as Arago writes, "He became sensible of his real vocation").

Educated in a military school directed by monks, Fourier wavered between a career in the church or the military. He preferred the latter, but this was not then possible because Fourier's father had been a tailor and not of the nobility. Fourier thus entered an abbey, but before taking his vows the social upheaval in France attracted him to a teaching position. He was appointed to the principal chair of mathematics in the Military School of Auxerre.

He soon displayed an unusual talent for lecturing on rhetoric, history, and philosophy, substituting for his colleagues when they became ill. His lectures on these several topics attracted a delighted audience of diverse backgrounds. This characteristic distinguished Fourier throughout his career.

In 1789 Fourier read a paper to the Paris Academy of Sciences on the resolution of numerical equations of all degrees. This work, enunciated by Fourier at the age of twenty-one, formed the cornerstone upon which he developed his future mathematical work. In this paper, Fourier extended some of the prior contributions of Lagrange. Nevertheless, Fourier's accomplishment had little appeal to the pure mathematicians, who felt that it lacked rigor. To the physical scientist, however, Fourier's results were received warmly (and still are) since they simplified calculations.

Returning to Auxerre, Fourier enthusiastically embraced the principles of the Revolution. This was a period of revolution not only politically but in the arts and sciences as well. For example, the reformation of weights and measures was begun at this time, leading to the introduction of the metric system. Alas, the times had taken many of the savants into military activity or, as with Lavoisier, had removed them permanently (Lavoisier was guillotined).

Fortunately, Napoleon in his rise to power realized the futility of ignorance in building a meaningful empire and he encouraged the creation of schools. In 1794, the Ecole Normale was started and Fourier was rewarded for his patriotism in Auxerre by being appointed to the chair of mathematics. This school lasted but a few months, at which time the Ecole Polytechnique was established. Again, Fourier was called and he responded by embellishing his reputation for clearness, method, and erudition. His lectures attracted a fastidious and wide audience.

Soon after the formation of the Ecole Polytechnique, Napoleon began to dream of restoring Egypt to its ancient splendor. And what better way to accomplish this than by introducing French culture to this now-backward country? Napoleon realized that to achieve this goal he would need more than a mere army. He would require leading scientists, and he chose Gaspard Monge and Claude Louis Berthollet. Both of these men were members of the Paris Academy of Sciences, on the faculty of the Ecole Polytechnique, and recognized as being among the leading scientists of the time. They in turn asked Fourier to join them and he did so. In Cairo they established the Institute of Egypt with Monge as the first president and Fourier as perpetual secretary.

In Egypt, Fourier distinguished himself by extending his mathematical researches to general solutions of algebraic equations, methods of elimination, and indeterminate analysis. His breadth of interest was also manifested by his contributions in general mechanics. He designed an aqueduct to conduct water from the Nile to the Castle of Cairo, espoused a proposal to explore the site of ancient Memphis, gave a descriptive account of the revolutions and manners of Egypt, and designed a wind machine to promote irrigation.

However, Napoleon's dream to rescue the Egyptians from the galling yoke under which they had groaned for ages... [and] to bestow upon them without delay all the benefits of European civilization (I quote from Arago's eulogy) was a failure. The uncivilized Egyptians failed to respond to the cultural feast proffered. Napoleon surreptitiously returned to France, as did Monge, leaving Fourier to cope. During Fourier's continued stay in Egypt, Napoleon conquered much of Europe and became the virtual ruler of France. But his fortunes in Egypt never proved successful and within three years of Napoleon's stealthy departure, Fourier and the others went back to France.

Upon his return to France, Fourier was named Prefect of the Department of l'Isère. Although this area was a hotbed of political dissension, Fourier, with great diplomatic skill, soon established harmony among the near-warring factions. The situation was brought to such a quiet state that Fourier could continue his efforts in mathematics and letters. From Grenoble, the principal city of Isère, Fourier wrote his *Théorie Mathématique de la Chaleur*. This was Fourier's outstanding scientific achievement.

Fourier's effort received a mixed reception. The pure mathematicians again pointed out the lack of rigor in his treatment. Pure mathematicians and mathematical physicists have nearly always been at odds, the former disdaining any treatment that avoids the scrutiny of rigid proof and the latter pleased to have a procedure to express the results of their observations.

Fourier recognized that any function whose graph displays a periodicity can be considered to be a sum of sinusoidal functions. That is

$$f(x) = \sum_n A_n \sin nx$$

(The purist may note that I have taken some liberties in expressing the Fourier series. Pshaw.)

This series is now known as a Fourier series. Its real value to optics, of course, is that it leads to an integral transform whereby a periodic function of space, for example, may be transformed to a periodic function of time. This means that the spectral characteristics of a radiating source may be separated by a Michelson interferometer to provide a time-varying signal in which time is not directly related to wavelength. Generally, the transformation is undertaken with a computer although, before computers, other means were employed. [*sic*].

Fourier had submitted his treatment of the conduction of heat, in which his series was fundamental, to the Paris Academy in 1811, for which he was awarded its mathematical prize in 1812. As noted, some reservations were expressed with the favorable judgment. However, Fourier never admitted the validity of this dissension, giving unmistakable evidence near the close of his life that he thought it still unjust by causing this memoir to be published in the Academy records without changing a single word!

This work gave a tremendous impetus to the research of his colleagues who considered the geological heat content, the temperature of celestial regions, and the effects of heat on biological growth. During this period, Napoleon's influence had blossomed and faded. In 1815, Napoleon escaped from Elba and made a triumphal march on Paris. Fourier had mixed reactions to this news. He left Grenoble for Lyons, where some of the royalty had assembled. They greeted Fourier coldly and doubted that Napoleon could have captured nearby Grenoble. Consequently, Fourier was told to return and protect the (already fallen) city. Fourier had barely left Lyons when he was arrested by Hussars and conducted to Napoleon's headquarters. Fourier explained that his duty compelled him to act as he had. Napoleon forgave Fourier, but did not endear himself when he told Fourier, "I have made you what you are."

Fourier was appointed Prefect of the Rhône and given the title of Count—promotions Fourier dared not reject. However, this appointment as Prefect lasted but a short time. Fourier returned to Paris with no income and no financial reserve. It was a turbulent time for many. Napoleon's career ended at Waterloo, and the Bourbons were restored to power in Paris.

Fourier applied for a federal pension for his 15 years of service to his country. He was rudely denied. However, a former student at the Ecole Polytechnique, on learning of Fourier's plight, enabled him to receive the directorship of the Bureau de la Statistique of the Seine.

The Academy of Sciences sought at its first opportunity to elect Fourier to the society. Political intrigue, sanctioned by Louis XVII, prevented anyone who had been associated with Napoleon from election to the Academy. (Arago noted, "In our country, the reign of absurdity does not last long.") A year later, in 1817, the Academy again unanimously nominated Fourier to a place in the section of physics. This time there was royal confirmation without difficulty.

Fourier was now able to spend the last years of his life in retirement and in the discharge of academic duties. He became eloquent in discoursing on those facets of life which he had experienced. There are those who find this type of eloquence somewhat boorish, rather than fascinating. They cite an incident in Fourier's later years as a case in point. Fourier was seated at a table together with some who were strangers to him. One, in particular, was identified as an older officer. To him, Fourier described in great detail the events of a battle that had taken place in Egypt, of which Fourier had some first-hand knowledge. Fourier concluded his recitation of the details of the battle by noting, complacently, that he felt his memory had served him correctly in recalling these events. His companion, who seemed to have been enthralled by this discourse, assured Fourier that his statements were accurate and based this judgment on the fact that he, too, had personal knowledge of the battle, having been head of the Grenadiers involved!

Although endowed with a sturdy constitution, Fourier had adopted the habit of wearing too much clothing. Thus, although he gave the appearance of corpulence he was, in fact, a quite slender man. He abided in a sterile, ovenlike environment, even keeping his windows closed in the heat of summer. Visitors found this to be annoying. As a result of this, Fourier developed an aneurysm of the heart. In the spring of 1830 he sustained a fall while descending some stairs. This aggravated his condition and, within two weeks, he died.

Fouriers's name is now used as an adjective describing an elegant method of handling several optical processes. Thus, we who work in the field of optics revere this man. He was undoubtedly unaware that the legacy he left behind was rich beyond all expectations.

Speaking of Fourier methods, Ronald Bracewell in the first chapter of his *The Hartley Transform* reminds us that "... as with all evolution, there are antecedents." He refers us back to Euler (1707–1783), and "... a case can be made that, as long ago as the second century, Claudius Ptolemaeus (a Greek of Alexandria) used the same basic ideas."

ABBREVIATED REFERENCES APPROPRIATE TO THIS INTRODUCTORY MATERIAL

H. J. J. Braddick, *Vibrations, Waves, and Diffraction*
R. Buckley, *Oscillations and Waves*
D. C. Champeny, *Fourier Transforms in Physics*
C. A. Culver, *Musical Acoustics*
D. S. Falk, D. R. Brill, and D. G. Stork, *Seeing the Light*
I. Grattan-Guinness, *Joseph Fourier 1768–1830*
J. Herivel, *Joseph Fourier—The Man and the Physicist*
R. A. Waldron, *Waves and Oscillations*
R. H. Webb, *Elementary Wave Optics*
Volume 45 of *Great Books of the Western World* contains a biographical note on Fourier.

TOPICS FOR FURTHER CONSIDERATION

1. In Figure 1.12 the young woman is holding a 48-inch stick, her left thumb at the 9-inch mark. What is the spatial period and the spatial frequency of the pickets? What are the spatial frequencies of the grooves on a phonograph record? (You remember those. Can you find some to use for this?) Do all records have the same groove spatial frequency and is the spatial frequency the same at all record radii? What is the spatial frequency and spatial angular frequency of the teeth on your comb?

TOPICS FOR FURTHER CONSIDERATION

2. Graphically, or mathematically, or mentally, convince yourself that a sinusoidal "wave" that is shifted, or is moving, in the positive x-direction will use a negative-signed shift term in the argument of the trig function.

3. Consult a general physics book and discover what physical property of individual air molecules is related to the fluctuating pressure in a sound wave as it impinges on your ear drum or on a microphone. Is it the displacement of the molecules? Speed? Momentum? What?

    Looking forward to #4, can you find some understandable expressions which give the rate at which energy is propagated by a sound wave?

4. There are similarities between sound waves and light waves. Otoscopes shine light on eardrums. When light falls on an eardrum we refer to the [light energy per second per square meter], i.e., Joules/sec·m², as the *irradiance*. Impinging on your eardrum, how is the pressure in a sound wave related to the sound wave's "irradiance"? [energy/sec·m²] (The dimensions of pressure are [force/area], in MKS units, Newtons/m².)

5. Consult a physics or electric technology book and discover what mechanism in a microphone converts a sound wave to an electrical wave. Assuming the sound wave is a perfect sinusoid, can this be done perfectly, i.e., without distorting the wave shape? Does the answer depend on the temporal frequency of the sound wave? (There are various microphone mechanisms for this conversion process, and one wonders if there are "new" mechanisms to be discovered among the "new" materials of the 21st century. Penn State researcher Kenji Uchino might suggest lead–lanthanum zirconate–titanate ceramic doped with tungsten oxide.)

6. Using a simple computer graphing program for testing, design a *periodic wave function* which can propagate in one dimension, the function being based upon elementary mathematical functions *other* than the trig functions. As a traveling periodic wave you will probably find that your function will have to be written with the argument $(kx \mp \omega t)$. For example, using the elementary function, *exponential*, could $y(x, t) = e^{(kx \mp \omega t)}$, sometimes written, $\exp(kx \mp \omega t)$, represent a periodic traveling wave? (For those familiar with it, notice there is no $i = \sqrt{-1}$ in the exponent.)

7.

    A. Sometime in your mathematical experience you are likely to learn that $\sin(\theta)$ can be written as an infinite series,

    $$\sin(\theta) = \theta - \frac{\theta^3}{3!} + \frac{\theta^5}{5!} - \frac{\theta^7}{7!} + \cdots \qquad (\theta \text{ in radians})$$

    Use your calculator (set in radian mode) to show that this series will indeed yield a good value for $\sin 60° = \sin(\pi/3 \text{ radians})$.

B. Similarly you will find that cos(θ) is given by

$$\cos(\theta) = 1 - \frac{\theta^2}{2!} + \frac{\theta^4}{4!} - \frac{\theta^6}{6!} + \cdots \qquad (\theta \text{ in radians})$$

Use your calculator (set in radian mode) to show that this series will indeed yield a good value for $\cos 60° = \cos(\pi/3 \text{ radians})$.

C. You will also learn that

$$e^\theta = 1 + \theta + \frac{\theta^2}{2!} + \frac{\theta^3}{3!} + \frac{\theta^4}{4!} + \cdots \qquad (\theta \text{ in radians})$$

With your calculator get the value of $e = e^\theta = e^1$.

D. It should be obvious that

$$e^{i\theta} = 1 + i\theta + \frac{(i\theta)^2}{2!} + \frac{(i\theta)^3}{3!} + \frac{(i\theta)^4}{4!} + \cdots \qquad (\theta \text{ in radians})$$

Show that

$$e^{\pm i\theta} = \cos(\theta) \pm i \sin(\theta)$$

One form of
The Euler Formula

2

FOURIER SERIES AND SPECTRA IN ONE-DIMENSION FOR FUNCTIONS OF FINITE PERIOD

In this chapter we will illustrate how one can Fourier analyze any reasonably well-behaved periodic function, $f(x)$. By well-behaved we mean:

1. $f(x)$ is single valued over its period.
2. $f(x)$ has a finite number of maximum and minimum values.
3. $f(x)$ has a finite number of discontinuities.
4. $\int |f(x)|\,dx$ over one period is finite.

These are the Dirichlet conditions and they guarantee that an infinite Fourier series will converge to the proper value of $f(x)$ where $f(x)$ is continuous, and at a jump in $f(x)$, the series will converge to the midpoint.

2.1 SOME DEFINITIONS OF FOURIER SERIES

Given a *periodic function of time*, $f = f(t)$, of *period* $2T$, we have a mathematical formula for the Fourier series which can be used to represent that function:

$$f(t) = a_0/2 + \sum_{n=1}^{n=\infty} [a_n \cos(n\pi t/T) + b_n \sin(n\pi t/T)]$$

where

$$a_n = (1/T) \int_c^{c+2T} f(t) \cos(n\pi t/T)\,dt \qquad (n = 0, 1, 2, 3, \ldots)$$

and

$$b_n = (1/T) \int_c^{c+2T} f(t) \sin(n\pi t/T) \, dt \quad (n = 1, 2, 3, \ldots)$$

There is no b_0.

For the *periodic function in the space domain*, $f(x)$, of *period 2L*, the formulas are as follows:

$$f(x) = a_0/2 + \sum_{n=1}^{n=\infty} [a_n \cos(n\pi x/L) + b_n \sin(n\pi x/L)] \quad (2.1)$$

where

$$a_n = (1/L) \int_c^{c+2L} f(x) \cos(n\pi x/L) \, dx \quad (n = 0, 1, 2, 3, \ldots) \quad (2.2)$$

and

$$b_n = (1/L) \int_c^{c+2L} f(x) \sin(n\pi x/L) \, dx \quad (n = 1, 2, 3, \ldots) \quad (2.3)$$

Let us try to restrict our discussion, henceforth, mainly to the space domain. The temporal frequency, f, will not show up in the space domain, so there should be no confusion with $f(x)$ meaning "function of x." Knowing something about $f(x)$, our job, if we are doing analysis, is usually to find all the a_n and b_n. It is rather standard in these formulas that $2L$ is taken to be the period of the function.

The above form is that used in the *Mathematical Handbook of Formulas and Tables* by Murray R. Spiegel, Schaum's Outline Series, McGraw-Hill Book Company. Very often these formulas quickly simplify when $f(x)$ can be written: as an even function, in which case all $b_n = 0$; or as an odd function, in which case all $a_n = 0$. In both cases the remaining integral further simplifies due to symmetry, and due to the fact that c can be chosen as zero, or as $-L$, whichever might be convenient.

IMPORTANT! *Some writers use different but equivalent expressions. If one attempts to use functions of this type one must be consistent with the particular reference in use.*

There is a second very useful form of the Fourier series, an exponential form. Because it utilizes $\sqrt{-1} = i$, from the mathematics of complex numbers, we often refer to this form as the complex Fourier series. The preceding and following relationships are usually developed in more detail in undergraduate college level mathematical methods books in sections dealing with Fourier series; see the references at the end of this chapter. For conversion from trig functions to exponential functions the Euler "identities" are used. The identity which seems to occur most often is

$$\sin\theta = \frac{e^{i\theta} - e^{-i\theta}}{2i} \qquad (2.4)$$

The other identity is

$$\cos\theta = \frac{e^{i\theta} + e^{-i\theta}}{2} \qquad (2.5)$$

A combination of these two easily yields,

$$e^{\pm i\theta} = \cos\theta \pm i\sin\theta \qquad (2.6)$$

Richard Feynman has been reported as referring to this last equation as "... this amazing jewel ... the most remarkable formula in mathematics." ... I'm convinced.

Using the Euler identities one can form the complex, or exponential, form for the Fourier series representation of a *periodic function* in space as

$$f(x) = \sum_{n=-\infty}^{n=+\infty} c_n e^{in\pi x/L} \qquad (2.7)$$

where

$$c_n = (1/2L) \int_c^{c+2L} f(x) e^{-in\pi x/L} \, dx \qquad (2.8)$$

Note that in the previous trigonometric formula the sum ran from one to infinity, $\sum_1^\infty$, while here we have a sum throughout the range $\pm\infty$, using $c_{\pm n}$. There is a simple relationship between the c_{+n} and c_{-n}:

$$\text{For real } f(x), \ c_{-n} = c_n^* \qquad (c_n^* = \text{complex conjugate of } c_n) \qquad (2.9)$$

Furthermore the c_n are related to the a_n and b_n:

$$c_{+n} = (a_n - ib_n)/2$$
$$c_{-n} = (a_n + ib_n)/2 \quad (2.10)$$
$$c_0 = a_0/2$$

$$a_n = c_{+n} + c_{-n} = 2\,\text{Re}\{c_n\}$$
$$b_n = i(c_{+n} - c_{-n}) = 2\,\text{Im}\{c_n\} \quad (2.11)$$
$$a_0 = 2c_0$$

Mathematical operations with exponentials are relatively easier than with the rather bulky trig functions. However, the Fourier series one calculates with exponentials may require conversion again from exponential back to trig functions in order for one to "see" the individual terms correctly. Again we caution the reader that some writers use different but equivalent expressions. Because of the physical nature of this book's later material we prefer these specifically physical expressions, using $2L$ as the period, written for the space domain.

Given a one-dimensional periodic function ("wave") in the space domain, it is possible to break it up (analyze it) into its Fourier components (series terms, harmonics, spectral components) by finding the a_n and b_n or c_n. Inversely we could synthesize the function by putting the proper spectral components together. Such analysis is often called spectrum analysis whether dealing with light or sound, or electronic signals, and the spectrum obtained this way is often called a Fourier spectrum.

One further observation: for the Fourier series described above, sine terms, cosine terms, or exponential terms, the *constant term* is always the average value of $f(x)$ over one period. $a_0/2$ or c_0 are this average.

2.2 TRIGONOMETRIC REPRESENTATION OF FOURIER SERIES

Let us illustrate the manner in which such spectra could be displayed. Consider the square wave shown in Figure 2.1, and the relationships expressed in Chapter 1, $k = 2\pi s = 2\pi/\lambda$. Note that in our space domain Fourier series notation, λ of our periodic functions will always be called $2L$, so

$$k = 2\pi s = 2\pi/2L.$$

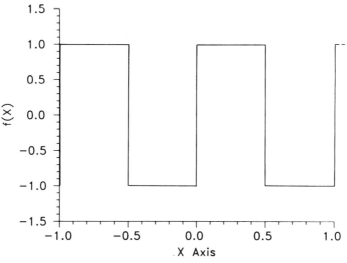

Figure 2.1 A square wave.

For the square wave of Figure 2.1 you can see that its spatial period, $\lambda = 2L = 1.0$, perhaps one meter; or we could say, 1.0 per cycle, perhaps one meter per cycle. Its spatial frequency s is $1/2L$ which is 1.0 cycle per unit length, perhaps one cycle per meter. Its spatial angular frequency k is $2\pi s$ which is 2π radians per unit length, perhaps per meter. For our formulas then, let's take $L = 0.5$ meter. It is an odd function, so all a_n will be zero.

The Fourier series for this wave is the infinite series with $n = 1, 3, 5, \ldots$:

$$f(x) = (4/\pi)\left[\frac{\sin(kx)}{1} + \frac{\sin(3kx)}{3} + \frac{\sin(5kx)}{5} + \cdots + \frac{\sin(nkx)}{n} + \cdots\right]$$

$$f(x) = (4/\pi)\left[\frac{\sin(2\pi sx)}{1} + \frac{\sin(6\pi sx)}{3} + \frac{\sin(10\pi sx)}{5} + \cdots + \frac{\sin(n2\pi sx)}{n} + \cdots\right]$$

$$f(x) = (4/\pi)\left[\frac{\sin(2\pi x/2L)}{1} + \frac{\sin(6\pi x/2L)}{3} + \frac{\sin(10\pi x/2L)}{5} + \cdots \right.$$
$$\left. + \frac{\sin(n2\pi x/2L)}{n} + \cdots\right]$$

where we have written $f(x)$ three different ways to show how it would appear in terms of all the quantities, spatial angular frequency, k, or spatial frequency, s, or period $2L$. The physical meaning is clearly seen. Consider the second option where the sine function arguments are written in terms of the fundamental spatial frequency s. The arguments could be written $2\pi(1s)x$, $2\pi(3s)x$, $2\pi(5s)x$, $2\pi(7s)x$, etc., the interpretation being that *the first sine term has a frequency*

28   FOURIER SERIES AND SPECTRA IN ONE-DIMENSION

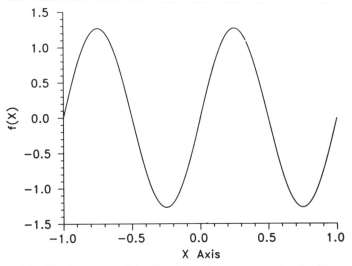

Figure 2.2 The first term of the Fourier series representation for Figure 2.1.

equal to that of the square wave itself 1s; *this is the fundamental frequency s.* (From above, $s = 1$ for this example.) It is often the dominant term, the term of largest amplitude; but not always. The frequency of the second sine term is $3s$, that of the third term, $5s$, and so on. Notice that the amplitude of each term is $4/\pi$ times 1, $\frac{1}{3}$, $\frac{1}{5}$, $\frac{1}{7}$, etc. Figure 2.2 is a plot of the first term of the series, Figure 2.3 is the second term, Figure 2.4 is the third, and Figure 2.5 is the sum

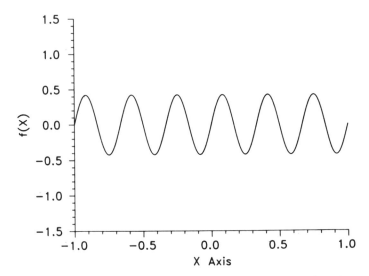

Figure 2.3 The second term of the Fourier series representation for Figure 2.1.

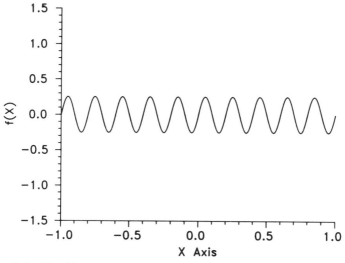

Figure 2.4 The third term of the Fourier series representation for Figure 2.1.

of the first three terms. Notice how the sum is beginning to look something like our square wave of Figure 2.1; and notice that within a half period, 0.5 meters, the fundamental term has one maximum, the second term has two maxima, the third term has three, and the sum has three. Do you see how just these three components of the square wave sum to zero where they should, $f(x) = 0$, and almost to one where they should?

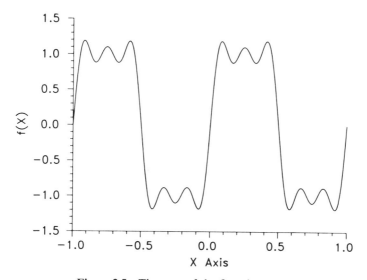

Figure 2.5 The sum of the first three terms.

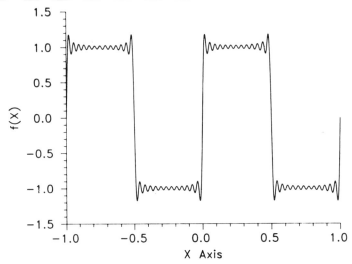

Figure 2.6 The sum of the first thirteen terms.

In Figure 2.6 we have added more terms of the series and you can guess quite correctly from the fact that within 0.5 meters there are thirteen maxima, that we used thirteen terms.

Reexamine the figures showing the individual waves and the ones showing the sums and see if you can answer the following: In reproducing the shape of the square wave what function is served by the fundamental and other low frequency terms, and what function is served by the high frequency terms? That is, what is responsible for the overall gross square wave structure, and what is responsible for sharpening up the edges and corners, the details, of the square wave?

You should see that in Figure 2.5 the overall gross structure of the square wave is present using only the first three (three lowest frequency) terms, whereas to flatten out the top of the wave and attempting to sharpen the corners requires the higher frequency terms. But even an infinite number of terms will not eliminate the "overshoot" seen at the function "corners" of Figure 2.6.

The overshoot in the Fourier series of $f(x)$ always occurs at the discontinuities of $f(x)$. This is often referred to as the "Gibb's overshoot" or Gibb's phenomena, after the physicist Gibb (1899), but the phenomena had been seen by Wilbrahan in 1848. The same phenomena occurs with electrical signals in time, when there is an abrupt (discontinuous) change in signal amplitude. The phenomena of ripples at the discontinuity is referred to as "ringing" and it clearly occurs on optical imaging with coherent light. [Walker, 1988, p. 61] The Gibb's overshoot is always 8.95% of the discontinuity amplitude. [Ramirez, 1985, p. 23].

Recall that earlier we mentioned the possibility of filtering out certain frequencies. We can do that here; Figure 2.7 is the same as the preceding one

TRIGONOMETRIC REPRESENTATION OF FOURIER SERIES   31

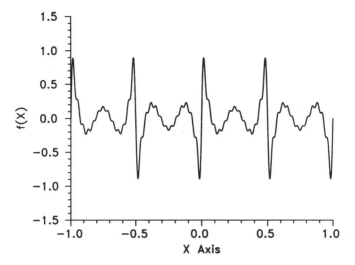

Figure 2.7 The sum of terms three through thirteen.

except missing the first two terms. As you would suspect, the fundamental is very important to synthesize the square wave, but you can see here the important role of the higher frequency terms in forming the edges and corners. To illustrate the filtering process further we will give one more example, and just remove the fundamental from the entire square wave; we plot only the one period from -0.5 meters to $+0.5$ meters (Figure 2.8). (In Figure 2.8 we simply

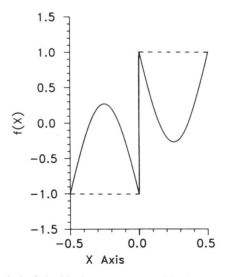

Figure 2.8 One period of the ideal square wave with the Fourier series fundamental, the first term, subtracted out.

subtracted the first term from the ±1 of the square wave in those two ±0.5 meter regions.) Without the fundamental term it doesn't look much like a square wave (dashed) does it? Later we will illustrate filtering in two dimensions, synthesizing without low frequency terms and similarities will be apparent between 2-dimensions and 1-dimension.

SOME EXAMPLES

We wish now to seemingly regress into some apparently very elementary examples, but by doing so we will illustrate some fundamental and important concepts.

Example 1

Write the Fourier series for a cosine wave. (Can't get more simple than that!) The wave is shown in Figure 2.9. The amplitude $= A$, period $= 2L$, spatial frequency $s = 1/\text{period} = 1/2L$. What makes this so simple is that the equation of the wave is immediately recognizable as its Fourier series which consists of a single term. The equation is,

$$f(x) = A \cos(\pi x/L)$$

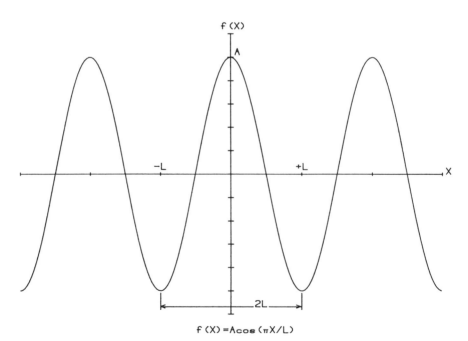

Figure 2.9 A cosine wave.

TRIGONOMETRIC REPRESENTATION OF FOURIER SERIES

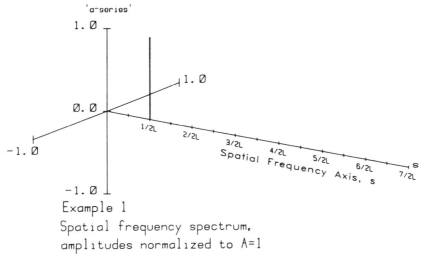

Example 1
Spatial frequency spectrum,
amplitudes normalized to A=1

Figure 2.10 The single spike at $s = 1/2L$ represents the amplitude of the single cosine term of Example 1.

The Fourier series for this wave,

$$f(x) = a_0/2 + \sum_{n=1}^{n=\infty} \{a_n \cos(n\pi x/L) + b_n \sin(n\pi x/L)\}$$

has $a_0 = 0$, $a_1 = A$. All other $a_n = 0$. Cosine is an even function so all $b_n = 0$. Thus only the a_1 term exists.

$$f(x) = a_1 \cos(n\pi x/L) = A \cos(\pi x/L)$$

Do you understand why the argument of the cosine function is $(\pi x/L)$? You can see that when $x = L/2$, $f(x) = 0$; and when $x = +L$, $f(x) = -A$.

And now we would like to try to represent the spectrum of this wave in terms of its spatial frequency, s, and its "strength," for which we will use $a_1 = 1$. In general, Fourier series are sums of many waves of many frequencies, so we will display frequency along one axis and the magnitude, a_1, along a second axis (Figure 2.10). Examine both figures of this example, the actual $f(x)$ and its spectrum $F(s)$. The spike represents a_1 at the appropriate spatial frequency, $s = 1/2L$.

Example 2

Let us just make up a finite series of cosine terms, add them together and look at the sum $= f(x)$, and then plot out the spectrum $F(s)$. Let us arbitrarily take:

Constant term:	$a_0/2 = A/2$	$s = 0$
next term	$(A/2)\cos(\pi x/L)$	$s = 1/2L$
next term	$(A/4)\cos(2\pi x/L)$	$s = 1/L$
next term	$-(A/2)\cos(3\pi x/L)$	$s = 3/2L$
next term	$(A/2\pi)\cos(5\pi x/L)$	$s = 5/2L$

Figure 2.11 shows what $f(x)$ = sum of this series looks like. We made it with only cosine terms, only a finite "a" series. To represent the spectrum $F(s)$ we plot out the amplitude of each term at each term's spatial frequency, s (Figure 2.12). There is a constant term also at $s = 0$, with amplitude $A/2$ set equal to $\frac{1}{2}$. Recall that the constant term, $A/2 = a_0/2$, represents the average value of $f(x)$ calculated over one period. Does it appear to be so in Figure 2.11? Even though this Fourier series was "dreamed up" the constant term is still the average value of the series; a fact made obvious when one recalls that the average value of sine and cosine terms is zero. The average value must derive from the constant term. All trigonometric Fourier series consist of, at most, one constant term and sine and cosine terms.

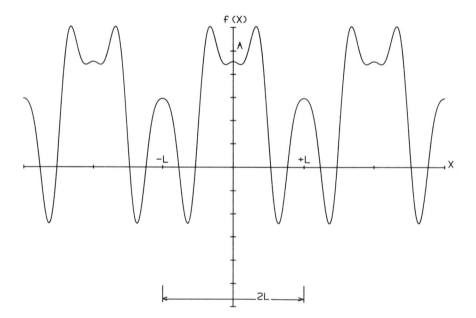

$f(X) = A/2 + (A/2)\cos(\pi X/L) + (A/4)\cos(2\pi X/L) - (A/2)\cos(3\pi X/L) + (A/2\pi)\cos(5\pi X/L)$

Figure 2.11 Example 2 has five cosine terms, all with a amplitudes.

TRIGONOMETRIC REPRESENTATION OF FOURIER SERIES

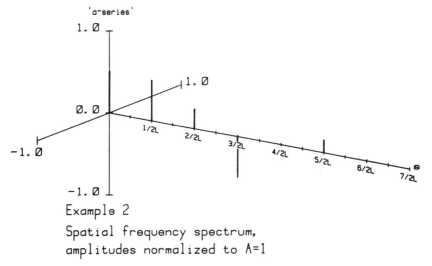

Example 2
Spatial frequency spectrum, amplitudes normalized to A=1

Figure 2.12 The five cosine terms, represented with proper amplitude spikes at their individual spatial frequencies.

Example 3

A Fourier series, in general, is composed of both cosine and sine terms. How shall we represent the sine terms, the "b_n series," so as not to confuse them with the "a_n series"? We can do this by recalling the relationship between cosine and sine functions,

$$\sin A = \cos(A - 90°) = \cos(A - \pi/2)$$

That is, a sine curve is identical to a cosine curve shifted 90° or $\pi/2$ radians. So in the spectrum diagram we represent the "b_n series" as spikes at 90° to both the "a_n series" and s-axes. The "$-90°$" tells us to orient the $+b_n$ spikes 90° clockwise looking down the s-axis toward the origin.

Then let us consider another completely arbitrary and finite Fourier series, this time with a "b_n series" also. We'll use the same terms from Example 2 and include some sine terms:

Constant term:	$a_0/2 = A/2$	$s = 0$
:	$(A/2)\cos(\pi x/L)$	$s = 1/2L$
:	$(A/4)\cos(2\pi x/L)$	$s = 1/L$
:	$-(A/2)\cos(3\pi x/L)$	$s = 3/2L$
:	$(A/2\pi)\cos(5\pi x/L)$	$s = 5/2L$
:	$(A/3)\sin(\pi x/L)$	$s = 1/2L$
:	$-(A/\pi)\sin(4\pi x/L)$	$s = 4/2L = 2/L$
:	$(A/6)\sin(6\pi x/L)$	$s = 6/2L = 3/L$

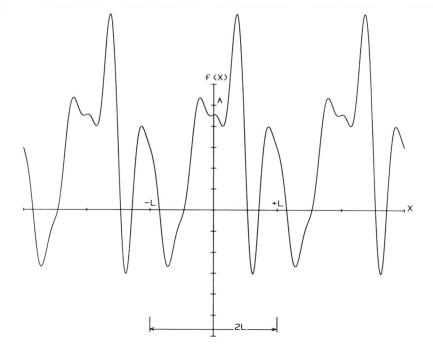

$f(X) = A/2 + (A/2)\cos(\pi X/L) + (A/4)\cos(2\pi X/L) - (A/2)\cos(3\pi X/L)$
$ + (A/2\pi)\cos(5\pi X/L) + (A/3)\sin(\pi X/L) - (A/\pi)\sin(4\pi X/L)$
$ + (A/6)\sin(6\pi X/L)$

Figure 2.13 This $f(x)$ contains both sine and cosine terms; its series will include both a_n and b_n terms.

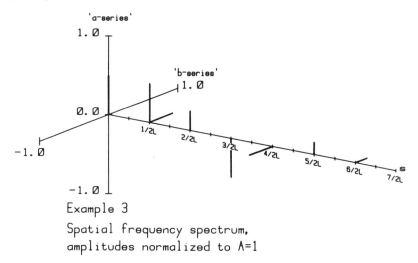

Example 3
Spatial frequency spectrum,
amplitudes normalized to A=1

Figure 2.14 The a amplitudes and the b amplitudes are plotted with respect to their own axes.

Again we will let $a_0/2 = A/2 = \frac{1}{2}$. The sum of these terms is shown in Figure 2.13 and the spectrum is in Figure 2.14, where we can now also see some b_n spikes.

Comment on adding Fourier series: It must be obvious to the reader that since the functions we have been dealing with *are* their Fourier series (which is a sum of terms) then if we were to add two functions together we could get the same result by summing the Fourier series of the two functions. Another way of saying it: if two functions are added together, the Fourier spectrum of the sum is the sum of their individual spectra. This is called the linearity property of Fourier series; it is quite useful as we shall have occasion to illustrate.

2.3 EXPONENTIAL, OR COMPLEX REPRESENTATION, OF FINITE FOURIER SERIES

Example 4

We repeat Example 1 in exponential form. Example 1 was,

$$f(x) = A \cos(\pi x/L)$$

a simple cosine wave, amplitude A, period $2L$, frequency $s = 1/2L$. According to the formula for exponential series, (2.7),

$$f(x) = \sum_{n=-\infty}^{n=+\infty} c_n e^{in\pi x/L}$$

$$c_n = (1/2L) \int_c^{c+2L} f(x) e^{-in\pi x/L} \, dx \qquad \text{and for real } f(x), c_{-n} = c_n^*.$$

The Fourier series in exponential form is not too obvious so we will carry out the calculations for this very simple example. The quantity c_0 is the constant term.

$$c_0 = (1/2L) \int_{-L}^{+L} A \cos(\pi x/L) e^{-in\pi x/L} \, dx, \qquad \text{and } n = 0 \text{ for } c_0$$

$$c_0 = (A/2L) \int_{-L}^{+L} \cos(\pi x/L) \cdot 1 \, dx = 0$$

The average value of a cosine is zero isn't it?

Solving for c_n in general we find that for all $n \geqslant 2$ all c_n and c_{-n} are zero. The general approach leads to an indefinite form $(0/0)$ when $n = +1$. So we

38 FOURIER SERIES AND SPECTRA IN ONE-DIMENSION

will solve specifically for c_{-1} and c_{+1}.

$$c_1 = (A/2L) \int_{-L}^{+L} \cos(\pi x/L) e^{-i\pi x/L} \, dx$$

We'll use the Euler identity to replace the exponential,

$$c_1 = (A/2L) \int_{-L}^{+L} \cos(\pi x/L) \cdot [\cos(\pi x/L) - i \sin(\pi x/L)] \, dx$$

Multiplying this out and solving by the use of integral tables we find

$$c_1 = A/2, \quad \text{and} \quad c_{-1} = c_1^* = A/2, \text{ also.}$$

Therefore,

$$f(x) = \sum_{n=-\infty}^{n=+\infty} c_n e^{in\pi x/L}$$

$$f(x) = c_{-1} e^{-i\pi x/L} + c_1 e^{+i\pi x/L} = c_1 e^{+i\pi x/L} + c_{-1} e^{-i\pi x/L}$$

With $c_1 = c_{-1} = A/2$,

$$\boxed{f(x) = (A/2)[e^{i\pi x/L} + e^{-i\pi x/L}]}$$

which has a somewhat different appearance than

$$f(x) = A \cos(\pi x/L)$$

but in reality, the two forms are identical. We identify $\theta = \pi x/L$ and we immediately see, through the Euler identities,

$$f(x) = (A/2)[e^{i\pi x/L} + e^{-i\pi x/L}] = (A/2) \cdot 2 \cos(\pi x/L) = A \cos(\pi x/L)$$

Exponential mathematics is often more simple than trigonometry for which reason it is often employed in spectrum analysis.

In the exponential form

$$f(x) = (A/2)[e^{i\pi x/L} + e^{-i\pi x/L}]$$

we see that the spectrum has two components now, both of amplitude $+A/2$. The trig function argument is $2\pi sx$, and here, $2\pi sx = \pi x/L$, meaning $s = 1/2L$. Thus one spike is at frequency $s = 1/2L$, the other at frequency $s = -1/2L$. (A negative frequency? We will discuss it shortly.) Figure 2.15 is the cosine wave. Figure 2.16 is its exponential, or complex, spectrum using A

EXPONENTIAL, OR COMPLEX REPRESENTATION 39

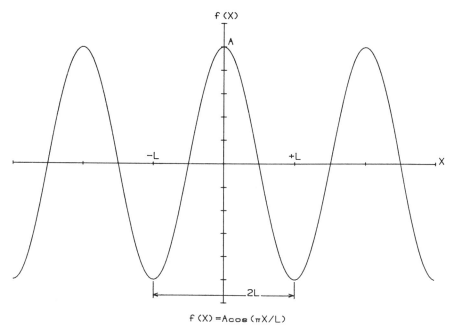

$f(X) = A\cos(\pi X/L)$

Figure 2.15 This cosine wave will be represented with complex notation.

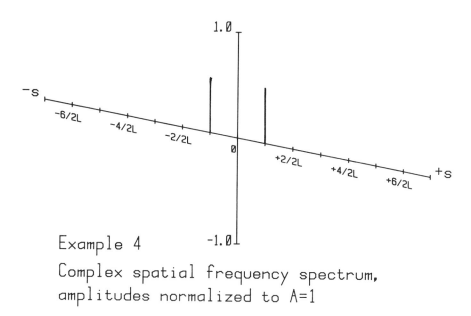

Example 4
Complex spatial frequency spectrum,
amplitudes normalized to A=1

Figure 2.16 In complex Fourier series notation the series has two terms, one at a "negative" spatial frequency.

equal to 1. Note the component with negative frequency.

Example 5

What will the *exponential* Fourier series look like for the function,

$$f(x) = A \cos(\pi x/L) + B \sin(2\pi x/L)?$$

Figure 2.17 is its plot, with $A = 1.0$, $B = 0.6$, $L = $ a constant.

Instead of using the Fourier formula or the identities between c_n, a_n, and b_n, let's just use the Euler relationships, (2.4) and (2.5):

$$\sin \theta = \frac{e^{i\theta} - e^{-i\theta}}{2i} \quad \text{and} \quad \cos \theta = \frac{e^{i\theta} + e^{-i\theta}}{2}$$

Being careful to identify the θ's correctly,

$$f(x) = A \frac{e^{i\pi x/L} + e^{-i\pi x/L}}{2} + B \frac{e^{i2\pi x/L} - e^{-i2\pi x/L}}{2i}$$

Multiplying the second term by i/i, and noting that $i^2 = -1$,

$$f(x) = (A/2)[e^{+i\pi x/L} + e^{-i\pi x/L}] + (B/2)[ie^{-i2\pi x/L} - ie^{+i2\pi x/L}]$$

The second term has i factors indicating that there will be both real and imaginary parts to this total spectral representation. In the spectrum plot we label the vertical axis "Real" and perpendicular to it an axis of

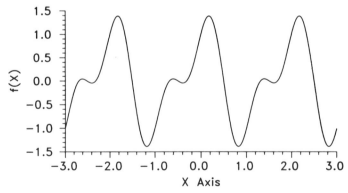

Figure 2.17 Example 5: In trig notation this $f(x)$ has both a cosine and a sine term. What will its complex spectrum look like?

"Imaginary" quantities; the spatial frequency axis, as before, is perpendicular to both.

The spatial frequencies are contained in the exponentials. The numeric multiplier of ix is always $k = 2\pi s = 2\pi/2L$. For the total A term, $k = \pm \pi/L = 2\pi s$. For the A term $s = \pm 1/2L$, and the amplitudes of both parts of the A term are positive, see Figure 2.18.

For the B term, $k = \pm 2\pi/L = 2\pi s$. For the B term, $s = \pm 1/L$, and note that the amplitude of the $s = -1/L$ term is positive, $+Bi/2$, while the amplitude of the $s = +1/L$ term is negative, $-Bi/2$. We plot the spectral spikes so, parallel to the "Imaginary axis" (Figure 2.18). From Figure 2.17 we see that if $A = 1$, then $B = 0.6$; those relative values were kept in the spectrum plot (Figure 2.18).

This complex method has led to the employment of the concept of negative frequencies. Negative frequencies in these applications are completely fictitious; they are artificial and spring forth when using exponential notation to describe harmonic motion or wave motion. Proper use of the Euler relationship, as we showed above, reveals the real nature of the wave in visual trigonometric form. Indeed, if you convert an exponential solution with both its $\pm$ frequencies, over to a real sinusoidal function, the "negative frequencies" vanish. Though perhaps not obvious here in these simple examples, the complex method (exponential mathematics) is generally easier to use.

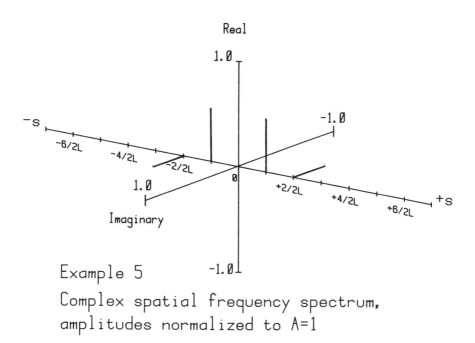

Figure 2.18 Two components contain i as an amplitude factor. We need an axis for such imaginary terms, with both positive and negative spatial frequencies.

Example 6

Let us now simply plot out a more complete example. We will just go back and use Example 3 which had eight terms including the $s = 0$ term. Since we already have the terms in trigonometric form we will simply use the Euler relations to write the complex terms. The terms with $e^{\pm i\pi x/L}$, where $s = 1/2L$ are taken as the fundamentals since they are of lowest frequency. The $\pm$ in the exponential also identifies the positive and negative frequency terms and hence also identifies the constants $c_{\pm n}$. In plotting the spectrum we will need to be careful for there are both cosine and sine terms, some with positive, some with negative amplitude, all to be translated as exponentials. Here we go, Example 6, from Example 3. Figure 2.13 is a plot of this function.

Frequency	Trigonometric	Exponential
$s = 0$	$A/2$	$A/2$
$s = 1/2L$	$(A/2)\cos(\pi x/L)$	$(A/2)[e^{i\pi x/L} + e^{-i\pi x/L}]/2$
$s = 1/L$	$(A/4)\cos(2\pi x/L)$	$(A/4)[e^{i2\pi x/L} + e^{-i2\pi x/L}]/2$
$s = 3/2L$	$-(A/2)\cos(3\pi x/L)$	$-(A/2)[e^{i3\pi x/L} + e^{-i3\pi x/L}]/2$
$s = 5/2L$	$(A/2\pi)\cos(5\pi x/L)$	$(A/2\pi)[e^{i5\pi x/L} + e^{-i5\pi x/L}]/2$
$s = 1/2L$	$(A/3)\sin(\pi x/L)$	$(A/3)[e^{i\pi x/L} - e^{-i\pi x/L}]/2i$
$s = 2/L$	$-(A/\pi)\sin(4\pi x/L)$	$-(A/\pi)[e^{i4\pi x/L} - e^{-i4\pi x/L}]/2i$
$s = 3/L$	$(A/6)\sin(6\pi x/L)$	$(A/6)[e^{i6\pi x/L} - e^{-i6\pi x/L}]/2i$

It will make our job easier if we get the i in the denominator of the last three terms into the numerator. So multiply those last three terms by i/i. The denominator $i^2 = -1$ will change the sign of each term. The last three terms become:

$$-(A/3)[ie^{i\pi x/L} - ie^{-i\pi x/L}]/2$$

$$+(A/\pi)[ie^{i4\pi x/L} - ie^{-i4\pi x/L}]/2$$

$$-(A/6)[ie^{i6\pi x/L} - ie^{-i6\pi x/L}]/2$$

Because of the i factor in each term, we see that these should all be plotted parallel to the $\pm$ imaginary axes. The complex spectrum components of these last six terms are:

$$A/3 \text{ at } s = +1/2L \qquad +A/\pi \text{ at } s = +4/2L$$
$$A/3 \text{ at } s = -1/2L \qquad -A/\pi \text{ at } s = -4/2L$$
$$-A/6 \text{ at } s = +6/2L$$
$$+A/6 \text{ at } s = -6/2L$$

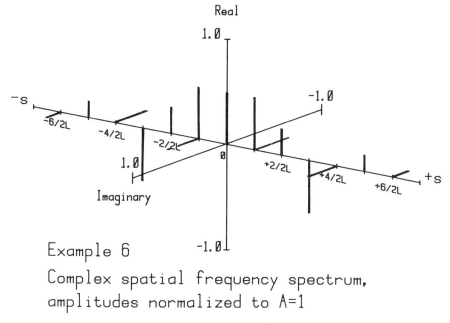

Example 6
Complex spatial frequency spectrum, amplitudes normalized to A=1

Figure 2.19 Example 6: The complex spectrum of spatial frequencies. There were seven trig terms, hence fourteen complex terms, and one zero frequency term.

There are fifteen terms altogether. The spectrum representation is shown in Figure 2.19. Now let's examine further the results of these examples.

2.4 TRIGONOMETRIC FOURIER SERIES AND SPECTRAL SPIKE ROTATION

Did you notice something about the Fourier spectrum plots using trigonometry, e.g., Example 3? If the "spike" for the cosine term, spatial frequency = 1/2L, part of the "a" series, is rotated about the s-axis, its projection back onto the "a-series/s-axis" plane represents the magnitude of that component wave *as a function of* x; a 2π rotation, one angular period, corresponds proportionately to an x-distance of $2L$, one spatial period, along the x-axis of this component wave. Figure 2.20 illustrates $\frac{1}{4}$ of a 2π rotation, a $\pi/2$ rotation corresponding to a $\frac{1}{4}$ of $2L$ increase in x, one-fourth the spatial period $2L$. A "spike" at the larger spatial frequency of $1/L$ (period = L) would do a complete 2π spike rotation in correspondence to a complete spatial distance of only L along the x-axis for this component wave. Corresponding to an advance in the positive x-direction it is conventional to visualize positive angular rotation as counter-clockwise viewed looking down the $+s$-axis toward the origin.

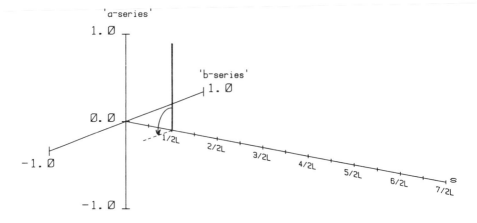

Spatial frequency spectrum, amplitudes normalized to A=1

Figure 2.20 If a particular position x yields a maximum value for an a_n term, then if we move $+\frac{1}{4}$ of a spatial period of that a_n term along the x axis of $f(x)$, it corresponds to a $\pi/2$ counterclockwise (positive) rotation of that spectral spike in the spectrum diagram.

Spikes corresponding to the "b-series" rotate with the same convention. Again the "b-spike" projection onto the "a-series/s-axis" plane, *the vertical plane*, represents the amplitude of that component wave *as a function of x*. All "b" terms are sine components; at $x = 0$ in the wave space, or angular position parallel to the "b" axis in spectrum space; they all contribute zero. Figure 2.21 shows the "b-term" positive rotation.

Clearly, if the nth spike rotates $+\Delta\theta_n$ this corresponds to a change in the amplitude of $f(x)$ brought about by an increased value of x, a new position along the x-axis. And it is easy to see that $\Delta\theta_n/\Delta x = k_n$, the angular spatial frequency [radians/meter] of that particular wave component. In the first example just cited, a 2π rotation corresponded to a $2L$ increase in x, and $k = 2\pi/2L = \pi/L$. In the second example, a 2π rotation corresponded to an L increase in x; $k = 2\pi/L$. The "spike" concept is a useful one for keeping tabs on all the spectral terms.

Nowhere in the latter discussion have we used the dimension of time. We have been dealing only with functions of x, functions in space. But with small effort you can see the extension of these ideas into wave motion, with proper consideration of the speed of each wave component. Also, if we wished to deal with functions of time, $f(t)$, we could set up an exact analogy, and relate ω_n to $\Delta\theta_n/\Delta t$.

Remember, the "spike" concept is not real. It is a mathematical construction, a visual concept which can prove helpful.

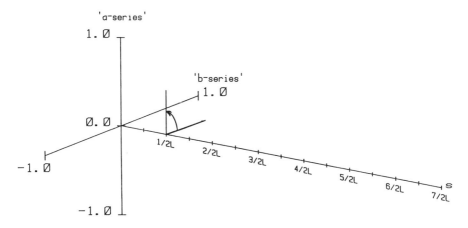

Spatial frequency spectrum, amplitudes normalized to A=1

Figure 2.21 A positive rotation of a b term spike at spatial frequency $1/2L$.

2.5 COMPLEX FOURIER SERIES AND SPECTRAL SPIKE ROTATION

Did you notice something about the complex Fourier series spectrum plots, with their positive and negative frequencies, and spikes only half the size of the trigonometric series? Here is the interpretation for the complex spectrum. Corresponding to increasing x values the positive frequency spikes rotate as before, positive in the counterclockwise direction, but the negative frequency spikes rotate clockwise (negative rotation) *both as viewed back toward the origin from the positive s direction.* Again the *real* magnitude of the wave is obtained from the spike projections onto the vertical plane. (How nice! That's the "real-axis/s-axis" plane!)

There are two spikes, however, for each frequency, $\pm s_n$. Consider the fundamental term, $s = 1/2L$, of Example 6,

$$(A/2)\cos(\pi x/L) = +(A/2)[e^{i\pi x/L} + e^{-i\pi x/L}]/2, \quad (n=1)$$
$$= +(A/4)[e^{i\pi x/L} + e^{-i\pi x/L}]$$

We see immediately that in the rotational sense this corresponds to

$$(A/4)[e^{+i\theta} + e^{-i\theta}]$$

$\theta = \theta_n = \theta_1$ here, and $\theta_1 = \pi x/L$. Again, as in the trigonometric portrayal, there

46 FOURIER SERIES AND SPECTRA IN ONE-DIMENSION

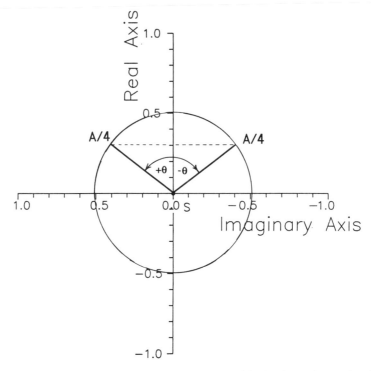

Figure 2.22 In a complex spectrum, as one moves positively along the x axis of $f(x)$, both positive and negative frequency spectral spikes rotate but in opposite directions. The positive frequency spike rotates positive (counterclockwise) while the negative frequency spike rotation is negative (clockwise).

is direct correlation between $\Delta\theta_n$ and an increased value of x to a new amplitude of $f(x)$ for each component wave.

The amplitude of the wave was $A/2$, but these "complex spikes" are both size $+A/4$, neither *term* is negative, but of course one is of negative frequency. See Figure 2.22 where the numerical axis has little meaning here. Summing the projections of the two $A/4$ spikes we have

$$(A/4)\cos(+\theta) + (A/4)\cos(-\theta) = (A/4)[\cos\theta + \cos\theta] = (A/2)\cos\theta$$

So both spikes are employed to get the amplitude of a harmonic of $f(x)$ corresponding to some $\Delta\theta_n$ and therefore some corresponding new increased value of x.

The second term of Example 5 was

$$+B\sin(2\pi x/L)$$

of amplitude B. Its complex form is

$$+(B/2i)[e^{i2\pi x/L} - e^{-i2\pi x/L}] = +(B/2i)\binom{i}{i}[e^{i2\pi x/L} - e^{-i2\pi x/L}]$$

$$-(B/2)[ie^{i2\pi x/L} - ie^{-i2\pi x/L}] = (-B/2)ie^{i2\pi x/L} + (B/2)ie^{-i2\pi x/L}$$

hence two spikes, one $-B/2$ at $s = +1/L$, one $+B/2$ at $s = -1/L$. They correspond to $(-B/2)[ie^{i\theta} - ie^{-i\theta}]$, see the spikes of Figure 2.23 with $\theta = 0$. The θ's indicated show how the spikes rotate as x increases in the space domain. Using the Euler relationship, $e^{\pm i\theta} = \cos\theta \pm i\sin\theta$, we have

$$(-B/2)[i(\cos\theta + i\sin\theta) - i(\cos\theta - i\sin\theta)]$$
$$= (-B/2)[i\cos\theta - \sin\theta - i\cos\theta - \sin\theta]$$
$$= +B\sin\theta \quad \text{or} \quad B\sin(2\pi x/L)$$

When $\theta = 0$, the sum of the $B/2$ spike projections are zero on the real-axis/s-axis plane as in Figure 2.23, but when $\theta = 90°$ we see we get the full amplitude B

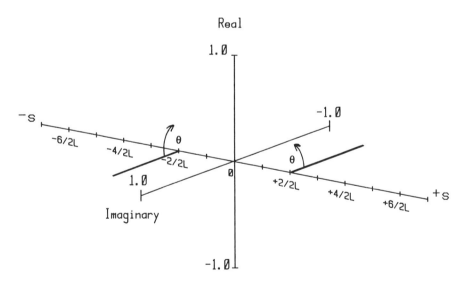

Complex spatial frequency spectrum, amplitudes normalized to A=1

Figure 2.23 As x increases in the space domain of $f(x)$, the positive and negative frequency terms of this complex spectrum rotate around the s axis, as shown.

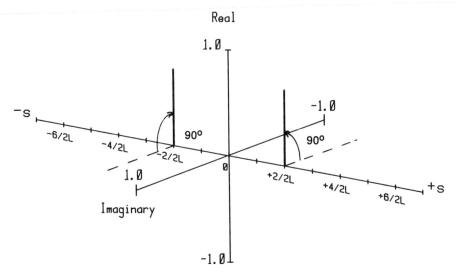

Figure 2.24 The complex spectrum of Figure 2.23 after a 90° rotation.

on the real axis, as in Figure 2.24, and similarly we would get the correct value at any other θ. So whether the complex spikes are related to the trigonometric "a-series" or "b-series," the sum of the two projections on the real-axis/s-axis plane will provide the correct amplitude for each harmonic as a function of rotational $\Delta\theta_n$, corresponding to an appropriate increase of x in the space domain to a new value of $f(x)$.

The preceding pages and illustrations should give the reader a fairly good idea how to calculate and visualize the spectrum of a Fourier series, both trigonometric and complex. It does, however, require hands-on practice to make sure you really understand.

2.6 A SHIFT IN $f(x)$—TRIGONOMETRIC CASE

What happens to the spectrum if $f(x)$ is not nicely positioned on the x-axis as either an even or odd function? Suppose $f(x)$ were just shifted a slight distance $+p$ along the x-axis. Consider again

$$f(x) = A\cos(\pi x/L) + B\sin(2\pi x/L)$$

A SHIFT IN f(x)—TRIGONOMETRIC CASE

Shifting the complete $f(x)$ a distance $+p$ means shifting both terms $+p$. $+p$-shifted $f(x)$ becomes

$$f(x) = A\cos[\pi(x-p)/L] + B\sin[2\pi(x-p)/L]$$
$$= A\cos(\pi x/L - \pi p/L) + B\sin(2\pi x/L - 2\pi p/L)$$

[You might note the mnemonic here, that the positive shift, $+p$, dictates the negative sign in $(x - p)$. You may recall that this carries over into the argument for wave motion, $f(x \pm vt)$. A positive traveling wave dictates the negative sign and vice versa.]

The spectrum of the above wave contains the same constituent waves as before the $+p$-shift. But obviously there is "something different." How will its spectrum appear? Figures 2.25 and 2.26 compare only the $+p$-shifted cosine to an unshifted one.

We previously showed (trigonometric treatment of Example 3) how to orient the spike for the $(A/3)\sin(\pi x/L)$, perpendicular to the a–s plane (corresponding to $-90°$ or $-\pi/2$). Now we have angles of $-\pi p/L$ for the cosine term and $-2\pi p/L$ for the sine term.

$$f(x) = A\cos(\pi x/L - \pi p/L) + B\sin(2\pi x/L - 2\pi p/L)$$

We thus rotate their spectral spikes that amount in the *negative rotation sense* (Figure 2.27).

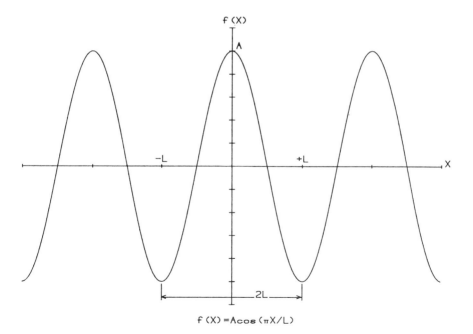

Figure 2.25 A cosine wave.

50   FOURIER SERIES AND SPECTRA IN ONE-DIMENSION

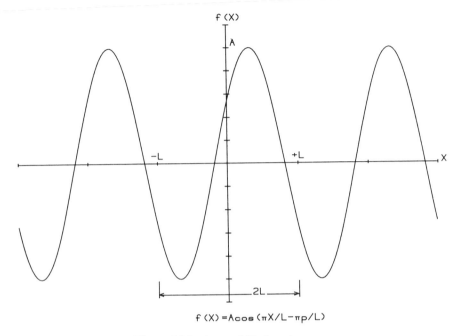

$f(X) = A\cos(\pi X/L - \pi p/L)$

Figure 2.26   A $+p$-shifted cosine wave.

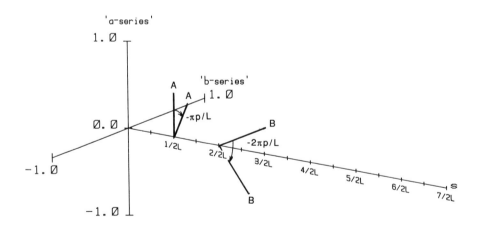

Spatial frequency spectrum, amplitudes normalized to A=1

Figure 2.27   In the spectrum of the trig representation, the additional phase term, e.g., $-\pi p/L$ is negative, hence negative rotation of the spectrum spike. We did not move in the $+x$ direction; $f(x)$ was shifted positively.

If you $+p$-shift a cosine term in the $+x$-direction, you will need to travel a distance in x equal to $+p$ to come to the cosine maxima that was originally at the origin. Hence the cosine spike (A) adjusted for $+p$ shift ($\phi = -\pi p/L$) needs to be rotated *positively* by $\pi p/L$ until it hits maximum projection on the a–s plane. When $x = +p$, $A\cos(\pi x/L - \pi p/L) = A$, the maximum.

Then note that the spike for the $+p$-shifted sine term (B) will be rotated $\phi = -2\pi p/L$ (Figure 2.27). If there were terms which prior to $+p$-shifting were: $3\pi x/L$, $4\pi x/L$, $5\pi x/L$, etc., their spikes would be rotated (for the $+p$-shifted $f(x)$): $-3\pi p/L$, $-4\pi p/L$, $-5\pi p/L$, etc. If we had some $f(x)$ which required an infinite Fourier series, ($n = 1$ to $n = \infty$) and we $+p$-shifted this $f(x)$, then each series term would employ $-n\pi p/L$, and each spike would be rotated $-n\pi p/L$ with respect to the spikes of the original unshifted $f(x)$. The quantity $-n\pi p/L$ is often called a *phase angle*. These graphs show how one might choose to picture them, if one could draw in 3-space.

2.7 A SHIFT IN $f(x)$—COMPLEX CASE

$f(x) = A\cos(\pi x/L) + B\sin(2\pi x/L)$ has complex form

$$f(x) = (A/2)[e^{+i\pi x/L} + e^{-i\pi x/L}] + (B/2)[ie^{-i2\pi x/L} - ie^{+i2\pi x/L}]$$

If $f(x)$ is $+p$-shifted, we again replace x with $(x - p)$ and find, for the complex notation,

$$f(x) = (A/2)[e^{+i(\pi x/L - \pi p/L)} + e^{-i(\pi x/L - \pi p/L)}]$$
$$+ (B/2)[ie^{-i(2\pi x/L - 2\pi p/L)} - ie^{+i(2\pi x/L - 2\pi p/L)}]$$

Taking the $(-)$ inside the exponential parentheses, and rearranging i's,

$$f(x) = (A/2)e^{+i\pi x/L - i\pi p/L} + (A/2)\underline{e^{-i\pi x/L + i\pi p/L}}$$
$$+ (iB/2)\underline{e^{-i2\pi x/L + i2\pi p/L}} - (iB/2)e^{+i2\pi x/L - i2\pi p/L}$$

we see immediately that the two *negative frequency components* (underlined above) are phase-shifted (broken underline) $+\pi p/L$ and $+2\pi p/L$, the *positive frequency components* (not underlined) are phase-shifted $-\pi p/L$ and $-2\pi p/L$. Hence the spectral spikes are rotated: the two negative frequency spikes are rotated $+\pi p/L$ and $+2\pi p/L$; the two positive frequency spikes are rotated $-\pi p/L$ and $-2\pi p/L$ (Figure 2.28).

If we had some $f(x)$ which required an infinite complex Fourier series ($n = -\infty$ to $n = +\infty$), and we $+p$-shifted this $f(x)$, then each series term would employ $\pm n\pi p/L$, and each complex spectral spike would be rotated $\pm n\pi p/L$ with respect to the spikes of the original unshifted $f(x)$.

52   FOURIER SERIES AND SPECTRA IN ONE-DIMENSION

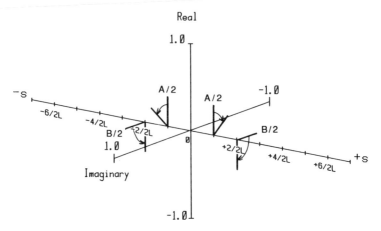

Complex spatial frequency spectrum, amplitudes normalized to A=1

Figure 2.28 In this complex spectrum example, the two negative frequency spikes have positive phase rotations, while the positive frequency spikes have a negative phase rotation.

If an $f(x)$ can be defined as even or odd, these phase terms and rotations are unnecessary. An even function will have a Fourier series of only cosine terms without phase shift; an odd function will have a series of only sine terms without phase shift. So if one is not interested in phase but simply in spectral spike frequency and amplitude, (i.e., the spectrum) it makes sense to make sure the $f(x)$ axis is positioned such that $f(x)$ can be defined as either even or odd.

2.8   VERTICAL SHIFTS

An odd $f(x)$ becomes neither even nor odd when lifted or lowered by a constant term $a_0/2$. It can be treated as an odd $f(x)$ and simply find $a_0/2$, its average value. a_0 affects neither the amplitude or frequency of the individual terms. $a_0/2$ is the $n = 0$, zero frequency term (sometimes called the dc term).

An even $f(x)$ remains even when lifted or lowered by a constant term $a_0/2$. Since the even $f(x)$ is a series of cosine terms, a_0 will be found along with all the other a_n.

Recognizing even or oddness can simplify calculation of the amplitudes of the Fourier series terms, the a_n, b_n, and/or c_n.

Some functions, $f(x)$, are neither even nor odd but can always be rewritten as a sum of an even and an odd function in the following manner. Given $f(x)$ neither even nor odd, rewrite it as follows:

$$f(x) = (\tfrac{1}{2})[f(x) + f(-x)] + (\tfrac{1}{2})[f(x) - f(-x)] \tag{2.12}$$

Of the two terms on the right side, the first is even, the second is odd. The Fourier series for $f(x)$ will thus become the sum of the Fourier series of one even function and one odd function—the linearity principle.

Some simplifying rules for even and odd $f(x)$:

$$\text{If } f(x) \text{ is odd} \begin{cases} b_n = (2/L) \int_0^L f(x) \sin(n\pi x/L) \, dx \\ a_n = 0 \end{cases}$$

$$\text{If } f(x) \text{ is even} \begin{cases} a_n = (2/L) \int_0^L f(x) \cos(n\pi x/L) \, dx \\ b_n = 0 \end{cases}$$

The average value of a function of period $2L = (1/2L) \int_c^{c+2L} f(x) \, dx$.

ABBREVIATED REFERENCES

G. Arfken (1970); M. Boas (1983); H. J. J. Braddick (1965); M. Cartwright (1988); R. Harding (1985); W. Kaplan (1952); E. Kreyszig (1972); R. K. Livesley (1989); M. C. Potter (1978); R. W. Ramirez (1985); K. F. Riley (1974); A. R. Shulman (1970); M. R. Spiegel (1968), (1971); E. G. Steward (1983); J. S. Walker (1988); C. W. Wong (1991); C. R. Wylie and L. C. Barrett (1982).

FOR THE READER TO TRY

1. For the odd $f(x)$ defined by:

$$f(x) = x, \quad \text{for } -\pi < x < \pi$$

take $2L$ as 2π, and use the formulas for a_n and b_n, (2.2) and (2.3), and show that the trig Fourier series, (2.1), for this $f(x)$ is

$$f(x) = 2\left(\frac{\sin x}{1} - \frac{\sin 2x}{2} + \frac{\sin 3x}{3} - \cdots\right)$$

Sketch $f(x)$. Then, if you have a computer that can plot $f(x)$, give it your solution to plot between $x = -3\pi$ and $x = +3\pi$, and see how many terms you might need to use to get good representation of this $f(x)$.

2. Using an elementary function, as we did above, make up your own periodic $f(x)$. Choose a nice period, $2L$, like $2L = 2$, find the Fourier series, have a computer plot many of the terms and see how well it reproduces your $f(x)$.

3. For the trig Fourier series you have composed, use your a_n and b_n and equations (2.10) and (2.9) to get the c_n. Compose the complex Fourier series using (2.7).

3

FOURIER SERIES AND SPECTRA FOR FUNCTIONS OF INFINITE PERIOD; ONE DIMENSIONAL

Now that we know how to Fourier analyze periodic functions we will, in this chapter, look at what happens when the period of a function, $2L$, approaches infinite extent; meaning, for practical purposes, a one-shot function, i.e., non-periodic. Letting the period become infinite is very similar to letting the period remain finite while shrinking the wave pulse width to zero.

3.1 DEALING WITH PERIOD AND WITH PULSE WIDTH FOR INFINITE SERIES

We will write the Fourier series for the wave shown in Figure 3.1, and then examine the results as we vary different aspects of the wave. We start with this nice mathematically simple even function. As an even function all the b_n will be zero. Using formula (2.2) for a_n, with $c = -L$ and $c + 2L = +L$, (the period is $2L$), the pulse width, d, is from -0.5 to $+0.5$,

$$a_n = (1/L) \int_c^{c+2L} f(x) \cos(n\pi x/L)\, dx = (1/L) \int_{-L}^{+L} f(x) \cos(n\pi x/L)\, dx$$

$$= (1/L) \int_{-d/2}^{+d/2} A \cos(n\pi x/L)\, dx$$

Doing the integration,

$$a_n = (2A/n\pi) \sin(n\pi d/2L), \qquad n \neq 0. \quad \text{(Why?)}$$

DEALING WITH PERIOD AND WITH PULSE WIDTH FOR INFINITE SERIES 55

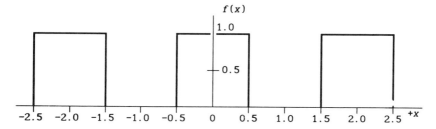

Figure 3.1 An always positive rectangular wave, the $f(x)$ for which we determine the Fourier series.

This $f(x)$ does have a non-zero average value, so a_0 is not equal to zero. You can see we need a separate calculation for $n = 0$; division by n equal zero is undefined. Use (2.2) again, with $n = 0$:

$$a_0 = (1/L) \int_{-d/2}^{+d/2} A\, dx = Ad/L$$

This was not obtainable from the general expression for a_n, above.

Because this $f(x)$ is even, all the $b_n = 0$. In addition to the $a_0/2$ term, which equals $Ad/2L$, the total $f(x)$ expression becomes

$$f(x) = Ad/2L + \sum_{n=1}^{n=\infty} a_n \cos\left(\frac{n\pi x}{L}\right)$$

$$f(x) = Ad/2L + \sum_{n=1}^{n=\infty} \underbrace{\left[\frac{2A}{n\pi}\sin\left(\frac{n\pi d}{2L}\right)\right]}_{a_n} \cos\left(\frac{n\pi x}{L}\right) \qquad (3.1)$$

It is important to notice that the amplitude, a_n, of each spectral term depends specifically on the ratio d/L. Also notice that a_n does not contain x. Regardless of the frequency of any term there is an envelope which shapes the amplitude of all spectral spikes; the envelope dictated by $\sin(n\pi d/2L)$. The frequency of each spectral spike, and the frequency spacing between spikes is dictated by L in $\cos(n\pi x/L)$.

Let $d/L = 1$ as in Figure 3.1.

$$f(x) = A/2 + (2A/\pi)[\cos(\pi x/L) - (\tfrac{1}{3})\cos(3\pi x/L) + (\tfrac{1}{5})\cos(5\pi x/L) - \cdots]$$

$$= A/2 + \sum_{n=1}^{n=\infty} \underbrace{\left\{(2A/\pi)(-1)^{n-1}\left(\frac{1}{2n-1}\right)\right\}}_{a_n} \cos[(2n-1)\pi x/L] \qquad (3.2)$$

The quantity in the curly brackets is a_n for this specific $f(x)$ with $d/L = 1$. The last expression will look better written in the following manner:

$$= A/2 + (2A/\pi) \sum_{n=1}^{n=\infty} (-1)^{n-1} \left(\frac{1}{2n-1}\right) \cos[(2n-1)\pi x/L]$$

The $a_{n(\text{even})}$ terms are missing due to the $\sin(n\pi/2)$ factor in (3.1), which for $d/L = 1$ generates only zeros and ones. The $(-1)^{n-1}$ alternates the sign ($\pm$) of each term; $(2n - 1)$ generates the 1, 3, 5, 7, ... sequence for the amplitude denominator and for the cosine argument.

The general term for $n \geq 1$ has the form

$$a_n \cos(n\pi x/L)$$

We would like to examine the frequency of each term and the frequency difference between terms; all contained in the cosine argument, $(n\pi x/L)$. a_n is simply the general amplitude of each term. A more general and standard notation for a cosinusoidal oscillation would be

$\cos(2\pi ft)$ in the time domain, but here,
$\cos(2\pi sx)$ in the space domain

where s is the spatial frequency. We want to compare this with our term, $\cos(n\pi x/L)$, so

$$n\pi x/L = 2\pi sx = 2\pi s_n x$$

Comparing the first and last, we see that for our series

$$s_n = n/2L \qquad (n = 1, 2, 3, 4, \ldots)$$

This gives us the spatial frequencies of each of our terms. With $d = L$, only the odd terms exist; the $a_{n(\text{even})}$ are all zero. The spatial frequencies needed to make up this wave are:

$s_1 = 1/2L$ (the fundamental	$s_2 = 2/2L$,	but $a_2 = 0$
$s_3 = 3/2L$	$s_4 = 4/2L$,	but $a_4 = 0$
$s_5 = 5/2L$	$s_6 = 6/2L$,	but $a_6 = 0$
$s_7 = 7/2L$, etc.		but $a_{\text{even}} = 0$

DEALING WITH PERIOD AND WITH PULSE WIDTH FOR INFINITE SERIES

The spatial frequency difference between terms is:

$$s_2 - s_1 = \Delta s_{21} = 2/2L - 1/2L = 1/2L$$
$$\Delta s_{32} = 3/2L - 2/2L = 1/2L$$
$$\Delta s_{43} = 4/2L - 3/2L = 1/2L$$
$$\Delta s_{54} = 1/2L, \qquad \text{etc., i.e., all the same, } 1/2L$$

For this specific d/L ratio = 1, the even n terms are missing, but $\Delta s_{31} = \Delta s_{53} = \Delta s_{75}$, etc., are also all the same, but equal to $1/L$.

It will be important to note that d, the pulse width, plays no role in determining s_n, the frequencies of terms, or Δs, the *frequency spacing*. Thus the spectrum of all rectangular pulses should have some distinctive similarities. And here would be a good place to plot out the spectrum of this series. You know what that will look like, and remember, with $d = L$ all $a_{n(even)} = 0$. See Figure 3.2, where the horizontal axis is n, not spatial frequency. For the amplitudes, a_n, which do exist, their value is given by (3.1)

$$a_{n(odd)} = \frac{2A}{n\pi} \sin\left(\frac{n\pi d}{2L}\right) \qquad A = 1, d/L = 1.$$

But we want now to establish a more general concept, relating the frequency of each term and the frequency difference between adjacent terms, for *any* d and L. (Recall that all the preceding had $d = L$.) In all cases, $2L$ is the fundamental period of our $f(x)$. Now, in general, for $n \geq 1$, a series term is given by (3.1),

$$\underbrace{\left[\frac{2A}{n\pi} \sin\left(\frac{n\pi d}{2L}\right)\right]}_{a_n} \cos\left(\frac{n\pi x}{L}\right)$$

Not being present in the cosine term, the quantity d (the pulse width) has no control over the frequency of the spectral terms or the frequency spacing. That is, d tells us nothing about *where* in a spectrum a spike will occur, nor their frequency spacing. What does control their frequency and frequency spacing? L controls both. Or one might say, $2L$, the period of $f(x)$ controls both. So again the frequencies will be $1/2L, 2/2L, 3/2L$, etc. And the frequency spacings will all be the same $1/2L$, whatever $2L$ happens to be. Whether an actual term at each frequency exists depends on a_n in the square brackets, i.e., on that sine factor; it periodically goes to zero and depends upon d/L. In general, the amplitude of each term gets smaller by $1/n$. We can still say $\Delta s = 1/2L$, but some a_n, and therefore some spikes, might be zero. Therefore, whatever $2L$ is, the spectral terms and spikes will start with a fundamental at $s_1 = 1/2L$ and then spacings, Δs, of $1/2L$.

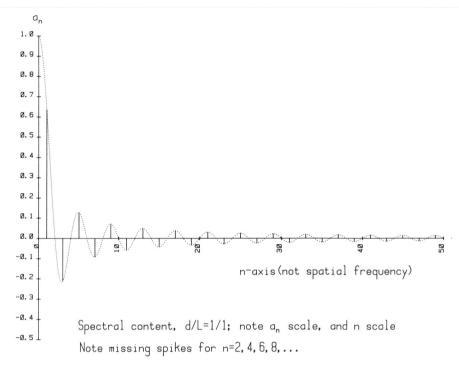

Figure 3.2 The spectrum; note the *n*-axis and a_n scale.

What does d (or nd/L) control? The amplitude of each term (spike) is governed by the ratio d/L in that sine factor and by $1/n$. If $d = L$, (one half the period)

$$a_n = (2A/n\pi) \sin(n\pi/2)$$

and the first zero a_n (and spike) occurs when $n = 2$ (when $n\pi/2 = \pi$) and there is only one spike prior to that at $n = 1$ (the fundamental). Of course there is one at s_0. But $s_2 = 2/2L = 1/L$; twice the fundamental, this is the first zero spike (this can all be seen in Figure 3.2).

Now if $d \neq L$ but $d \ll L$, examine $\sin(n\pi d/2L)$. How many n will occur before $n\pi d/2L = \pi$, where $a_n = 0$? Suppose $d/L = \frac{1}{100}$:

$$\text{then } n\pi d/2L = n\pi/200,$$

$$\text{and for this to be } = \pi, n = 200!$$

When $d = L$ we had one spectral spike before the first $a_n = 0$; now, with $d/L = \frac{1}{100}$, we will have 199 spikes, or series terms, before finally, $a_{200} = 0$ (see Figure 3.3). From the a_n scale note that each of the a_n are much smaller now that there are so many more contributing to the finite signal amplitude.

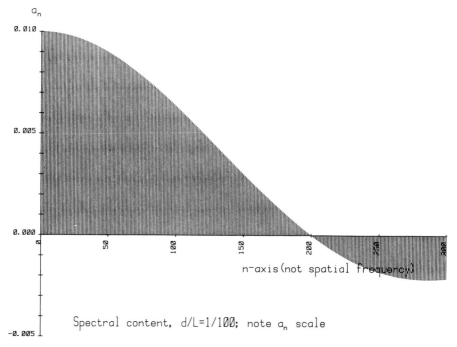

Figure 3.3 Spectral content with $d/L = \frac{1}{100}$; note a_n scale. [Note: the sequence of Figures 3.3–3.12 illustrates the change of spectral content as d changes, but $2L$ remains constant. Note the n-axis and the changing a_n scale.]

3.2 SIZE OF THE SPECTRAL SPIKES

When $d = L$, and $A = 1$,

$$a_1 = (2A/1\pi) \sin(1\pi d/2L) = (2A/\pi) \sin(\pi/2) = 2A/\pi$$

$$a_1 = 2/\pi = 0.637...$$

If $d = L$, the size of the a_1 spike is 0.637. See that this is so in Figure 3.2.

However, if $d/L = \frac{1}{100}$ what will be the size of a_1, the amplitude of the fundamental? For $d/L = \frac{1}{100}$, and $A = 1$,

$$a_1 = (2A/1\pi) \sin(1\pi d/2L) = (2A/\pi) \sin(\pi/200) = (2A/\pi)(0.0157)$$

$$a_1 = (2/\pi)(0.0157) = 0.01$$

See that this is so in Figure 3.3. For rectangular signals in the space domain, that sine factor (and of course the $1/n$ factor) shape the envelope of the a_n spikes.

60 FOURIER SERIES AND SPECTRA FOR FUNCTIONS OF INFINITE PERIOD

TABLE 3.1

d/L	a_1	$a_0 = Ad/L$	Figure
$\frac{1}{100} = 0.01$	0.0099	0.01	3.3
$\frac{1}{50} = 0.02$	0.0199	0.02	3.4
$\frac{1}{20} = 0.05$	0.0499	0.05	3.5
$\frac{1}{10} = 0.10$	0.0996	0.10	3.6
$\frac{1}{5} = 0.2$	0.197	0.20	3.7
$\frac{1}{2} = 0.5$	0.450	0.5	3.8
$\frac{1}{1} = 1.0$	0.637	1.0	3.9(3.2)
$\frac{3}{2} = 1.5$	0.450	1.5	3.10
$= 1.9$	0.0996	1.9	3.11
$= 1.99$	0.00999	1.99	3.12

If we keep $2L$ constant, but change d and hence d/L, we will need to change the scale of the a_n axis to make the a_n visible. So in the following plots, note that the a_n axis scale changes. Here then are some d/L ratios (Table 3.1) and plots (Figures 3.3–3.12) of the spectral content. In all cases $A = 1$, and the horizontal axis is n. With reference to the last line in Table 3.1: Remember when

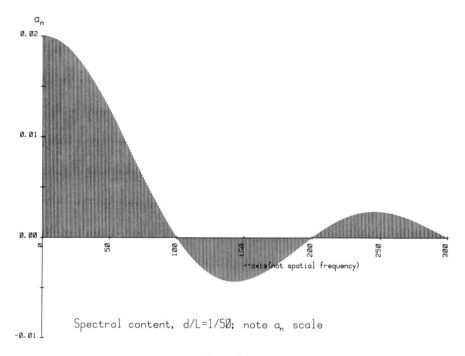

Figure 3.4

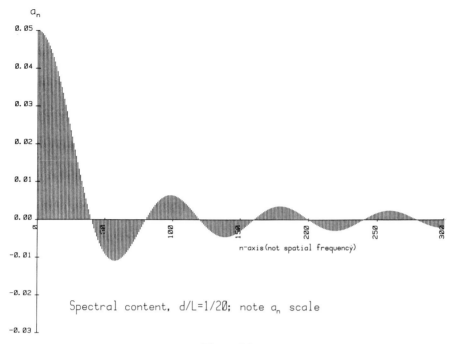

Figure 3.5

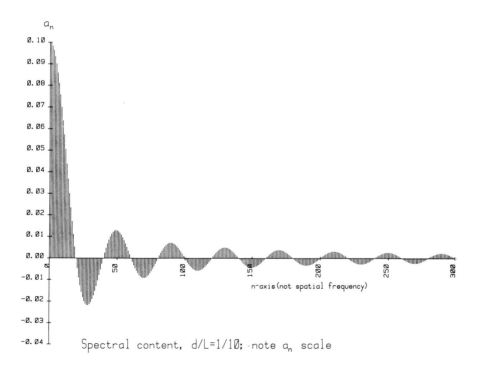

Figure 3.6

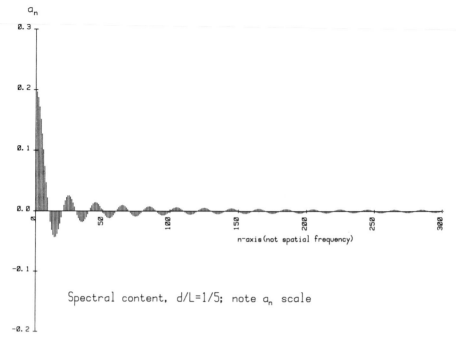

Figure 3.7

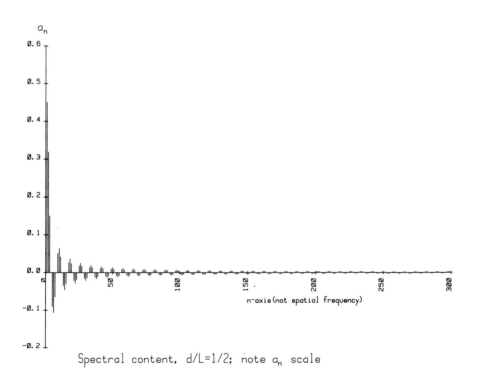

Figure 3.8

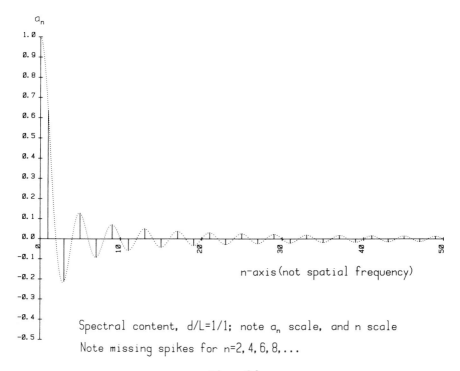

Figure 3.9

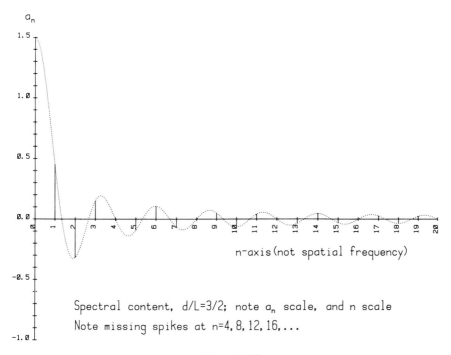

Figure 3.10

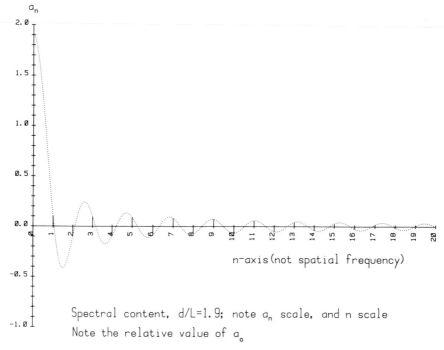

Spectral content, d/L=1.9; note a_n scale, and n scale
Note the relative value of a_o

Figure 3.11

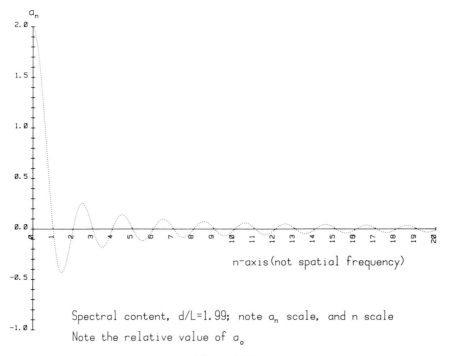

Spectral content, d/L=1.99; note a_n scale, and n scale
Note the relative value of a_o

Figure 3.12

$d/L = 2$, $d = 2L$, for our example, and $f(x)$ will equal a constant, A; all a_n are very small, insignificant, except a_0 which you can see is 1.99; $a_0/2$ must equal the average value, since A has been set equal to 1.0. It is coincidental, due to $A = 1$, that $d/L = a_0$.

The shape of our original $f(x)$, the always positive rectangular wave, that shape and size dictated the enveloping curve for the amplitude of all the spectral spikes just illustrated. Figure 3.13 is a composite summary of Figures 3.2 through 3.12.

3.3 LETTING THE PERIOD, 2L, APPROACH INFINITE EXTENT

Reexamine and note that as d/L became smaller, the number of spectral terms was increasing, while the frequency spacing, Δs, was decreasing. Examine the next set of Figures 3.14 through 3.19 where the horizontal axis is now s, spatial frequency, rather than n as before. So these are spectrum plots. Here we can consider d as constant, and $2L$, the period, constantly increases. As this happens, the spectral content progresses from a series of discrete terms to a continuum of harmonic content, that is, an infinite number of terms with essentially zero spacing in frequency. Look for this *inverse relationship* in all Fourier spectrum work: the *larger* the period of the function in the space domain, the *smaller* the spacing of the spectral components in the frequency domain.

Figure 3.20 is a summary of Figures 3.14 through 3.19. The a_n envelope again was dictated by the rectangular pulse shape; a_0 changes appropriately as d/L changes. You can see that since the horizontal axis is the same for all Figures 3.14 to 3.19, and d being constant, d must be the controlling quantity for the shape of the envelope, which does not change. Likewise the increasing $2L$ has dictated the shrinking Δs.

For this rectangular wave the general series term is:

$$\underbrace{\left[\frac{2A}{n\pi}\sin\left(\frac{n\pi d}{2L}\right)\right]}_{a_n}\cos\left(\frac{n\pi x}{L}\right)$$

a_n, in square brackets, is obtained from the Fourier formula by integration of $f(x)$. Hence a_n magnitudes are dictated by the shape of the pulse of $f(x)$. The $\cos(n\pi x/L)$ merely tells us the frequency of each spike.

Rearrange the factors in a_n, multiply and divide by d/L:

$$a_n = (d/L)A\,\frac{\sin[(d/L)n\pi/2]}{(d/L)n\pi/2}$$

66 FOURIER SERIES AND SPECTRA FOR FUNCTIONS OF INFINITE PERIOD

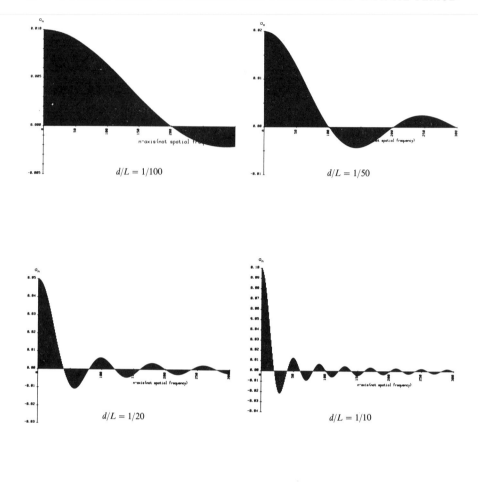

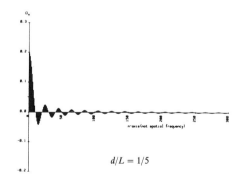

Figure 3.13 A comparison of Figures 3.3–3.12.

LETTING THE PERIOD, 2L, APPROACH INFINITE EXTENT

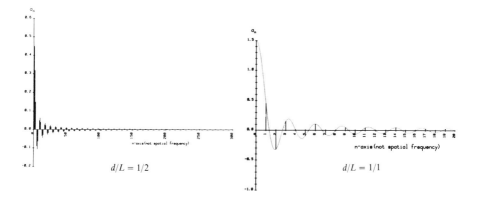

$d/L = 1/2$

$d/L = 1/1$

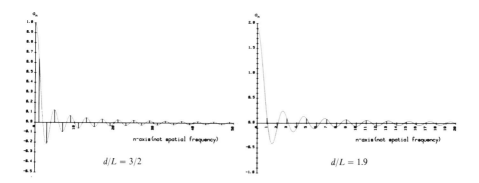

$d/L = 3/2$

$d/L = 1.9$

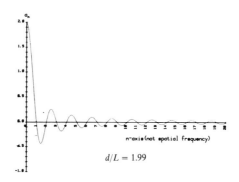

$d/L = 1.99$

Figure 3.13—*continued.*

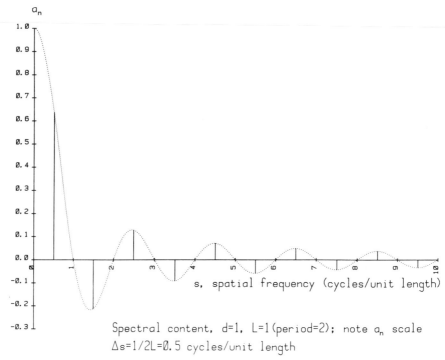

Spectral content, d=1, L=1(period=2); note a_n scale
$\Delta s = 1/2L = 0.5$ cycles/unit length

Figure 3.14 For the sequence of Figures 3.14–3.19, the horizontal axis is now spatial frequency. Pulse width $d = 1$ is kept constant as the period, $2L$, increases. These are spectrum plots.

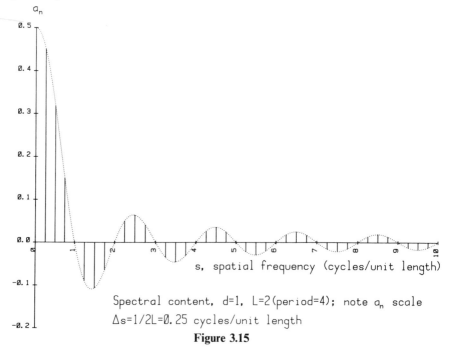

Spectral content, d=1, L=2(period=4); note a_n scale
$\Delta s = 1/2L = 0.25$ cycles/unit length

Figure 3.15

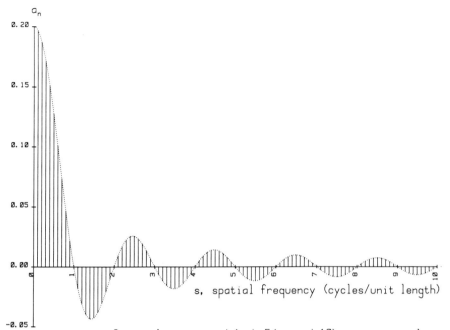

Spectral content, d=1, L=5(period=10); note a_n scale
$\Delta s = 1/2L = 0.1$ cycles/unit length

Figure 3.16

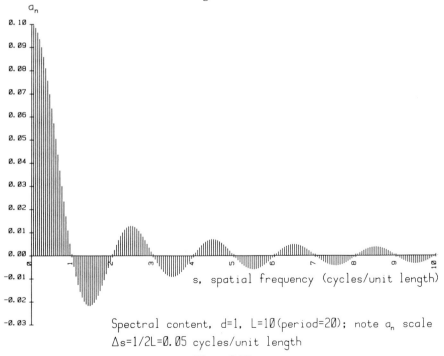

Spectral content, d=1, L=10(period=20); note a_n scale
$\Delta s = 1/2L = 0.05$ cycles/unit length

Figure 3.17

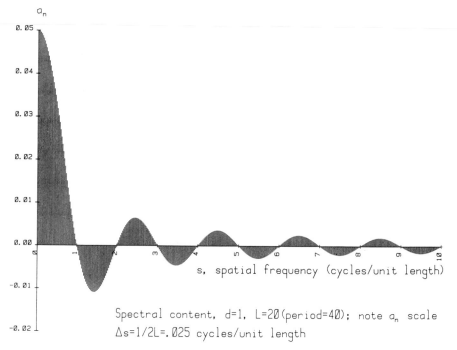

Spectral content, d=1, L=20(period=40); note a_n scale
$\Delta s=1/2L=.025$ cycles/unit length

Figure 3.18

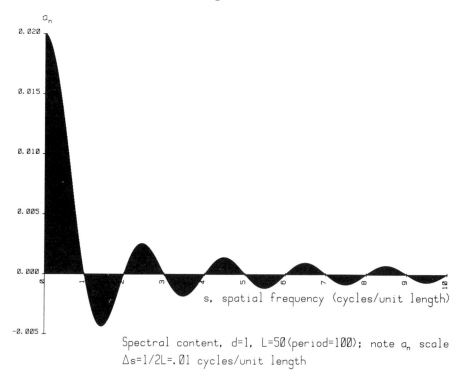

Spectral content, d=1, L=50(period=100); note a_n scale
$\Delta s=1/2L=.01$ cycles/unit length

Figure 3.19

As long as *d is a constant width of the rectangular pulse*, the spectral content envelope is of constant *shape* and it matters not what $2L$ is. If $d = 1$, then

$$a_n = (A/L) \frac{\sin(n\pi/2L)}{n\pi/2L}$$

The first zero of the envelope occurs when $(n\pi/2L) = \pi$. If L is large, say 100 (compared to $d = 1$), then the first zero will not occur until $n\pi/(2)(100) = \pi$, i.e., not until $n = 200$. With $d = 1$, this first zero always occurs at the same spatial frequency, the occurrences dictated from (3.1), the $\cos(n\pi x/L)$ factor, at values of s given by

$$n\pi/L = 2\pi s$$

$$200\pi/100 = 2\pi s$$

at the spatial frequency, $s = 1.0$ [cycles/unit length]. Furthermore, the sine function gives the oscillation of the envelope curve, while the $1/n$ factor decreases the amplitude as we step through each s_n.

From (3.1) note that the a_n

$$a_n = \frac{2A}{n\pi} \sin\left(\frac{n\pi d}{2L}\right)$$

could be written

$$a_n = 2A \frac{\sin(n\pi d/2L)}{n\pi} \cdot \frac{d/2L}{d/2L} = (Ad/L) \frac{\sin(n\pi d/2L)}{(n\pi d/2L)} \tag{3.3}$$

The cosine factor in (3.1) dictated the occurrences, $n\pi/L = 2\pi s$; i.e., occurrences at $s = n/2L$ (this is easy to see in Figure 3.15, but note that some $a_n = 0$). In our expression, (3.3), replace $n/2L$ with s, thus

$$a_n(s) = (Ad/L) \frac{\sin(\pi d s)}{(\pi d s)} \tag{3.4}$$

Hence, if we just look at the envelope of the $a_n(s)$ in these spectrum diagrams, it has the appearance of a continuous function. We shall use the expression $\text{sinc}(\beta)$ as a short form for $\sin(\beta)/\beta$. Thus we could write for the envelope function

$$a_n(s) = (Ad/L) \, \text{sinc}(\pi d s)$$

[**Caution:** Some authors prefer to define $\text{sinc}(\beta)$ as $\sin(\pi\beta)/\pi\beta$, e.g., Bracewell (1965). The $\text{sinc}(\pi d s)$ spectral envelope is common to rectangular $f(x)$.]

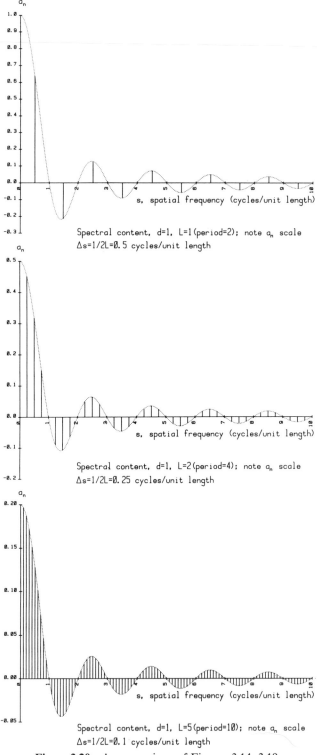

Figure 3.20 A comparison of Figures 3.14–3.19.

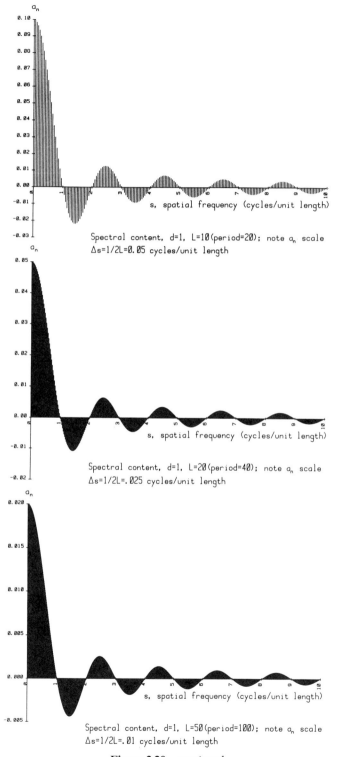

Figure 3.20—*continued.*

It is important to understand that the period, 2L, controls the frequency, s, of each term and Δs, the frequency spacing. **The envelope is controlled by d, the pulse width. If 2L, the period, gets large, Δs gets small.**

From Figure 3.13, if d gets large the spectrum envelope will keep its shape but shrink along the frequency axis toward the origin. As the period, 2L, increases, the spectrum changes from being a set of discrete terms which we have portrayed here as spikes of Δs spacing, see Figure 3.20. We see Δs approaches zero and indeed, if 2L becomes infinite, the spectrum is no longer a discrete set but a continuous function—an infinite set of wave components.

3.4 PHASE IN INFINITE FOURIER SERIES

The Fourier series for the even, always positive, rectangular wave was (3.2),

$$f(x) = \frac{A}{2} + \frac{2A}{\pi}\left[\frac{1}{1}\cos\left(\frac{\pi x}{L}\right) - \frac{1}{3}\cos\left(\frac{3\pi x}{L}\right) + \frac{1}{5}\cos\left(\frac{5\pi x}{L}\right) - \cdots\right]$$

When $d = L = 1$; period $= 2L = 2$; the $f(x)$, fundamental spatial frequency $= 1/\text{period} = \frac{1}{2} = s_1$; the specific wave ($f(x)$, a function of x, not s) can be written,

$$f(x) = \frac{A}{2} + \frac{2A}{\pi}\left[\frac{1}{1}\cos(2\pi s_1 x) - \frac{1}{3}\cos(2\pi 3 s_1 x) + \frac{1}{5}\cos(2\pi 5 s_1 x) - \cdots\right]$$

With $s_1 = \frac{1}{2}$ [cycles/unit length], the first term has a spatial frequency $\frac{1}{2}$; the second term, frequency $3s_1 = \frac{3}{2}$; the next term, frequency $5s_1 = \frac{5}{2}$; etc. The even number terms, as we explained, are missing. Figure 3.21 is a repeat of Figure 3.14 and, again with $A = 1$, you can see the proper amplitude spikes at their proper frequencies: $(2A/\pi) = +0.637$ at $\frac{1}{2}$. $(-2A/3\pi) = -0.212$ at $\frac{3}{2}$, $(2A/5\pi) = +0.127$ at $\frac{5}{2}$, $(-2A/7\pi) = -0.091$ at $\frac{7}{2}$, etc. The dotted envelope has the form sinc(πs); look where the maxima and minima occur. Let us redraw that spectrum, as far as to show the spike at $s = \frac{7}{2}$, (see Figure 3.22).

If $f(x)$ is lifted up or lowered, i.e., shifted vertically, the spectrum will not change but a_0 will change such that $a_0/2$ would be the new average value of $f(x)$. But what about horizontal shifts? How will they affect the spectrum?

3.4.1 Phase Shifts When $2L = 2$, $p = +\frac{1}{2}$

If this rectangular wave, $f(x)$, is shifted $+p$ along the x-axis, the spectral content won't change but all the series terms will need to be shifted in phase by the

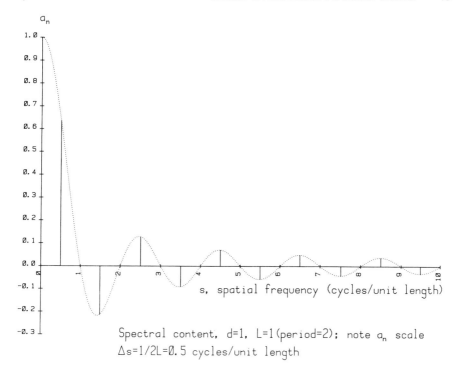

Spectral content, d=1, L=1 (period=2); note a_n scale
$\Delta s = 1/2L = 0.5$ cycles/unit length

Figure 3.21 Again, the spectrum when the pulse width $d = 1$, and period $2L = 2$.

proper amount. The series, rewritten for the $+p$-shifted wave, is

$$f(x - p) = \frac{A}{2} + \frac{2A}{\pi}\left[\frac{1}{1}\cos 2\pi s_1(x - p) - \frac{1}{3}\cos 2\pi 3 s_1(x - p) \right.$$
$$\left. + \frac{1}{5}\cos 2\pi 5 s_1(x - p) - \cdots\right]$$

If we let the physical shift in $f(x)$ be $+p = L/2$ (one quarter period), and with $L = 1$ and $s_1 = \frac{1}{2}$, $p = \frac{1}{2}$, see Figure 3.23, the phase shift of the terms become:

first: $2\pi s_1 p = 2\pi(\frac{1}{2})(\frac{1}{2}) = \pi/2$
second: $2\pi 3 s_1 p = 2\pi 3(\frac{1}{2})(\frac{1}{2}) = 3\pi/2$
third: $5\pi/2$,
fourth: $7\pi/2$, etc.

76 FOURIER SERIES AND SPECTRA FOR FUNCTIONS OF INFINITE PERIOD

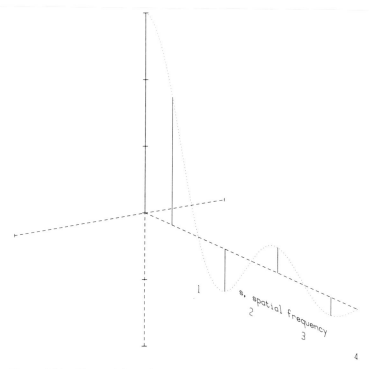

Figure 3.22 Figure 3.21 redrawn only out to spatial frequency $s = 4$.

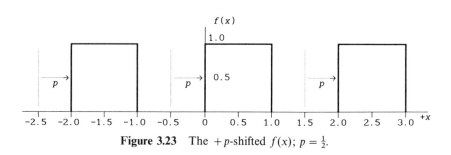

Figure 3.23 The $+p$-shifted $f(x)$; $p = \frac{1}{2}$.

The phase shift is $n\pi/2$ (note: only spikes at $n =$ odd exist). In Figure 3.24, sighting up the spatial frequency axis toward the origin, the spike of the first term ($n = 1$) has been rotated counterclockwise (the positive sense) through an angle $+\pi/2$. The second spike ($n = 3$), originally pointing down, is rotated $+3\pi/2$; the third ($n = 5$) is rotated $+5\pi/2$; the fourth ($n = 7$), $+7\pi/2$. The solid line marks the envelope of the spectral spikes for this $+p$-shifted $f(x)$; it is the loci of the tips of the spikes. If there were more spikes perhaps one might be better able to visualize the phase shifted spectrum.

PHASE IN INFINITE FOURIER SERIES

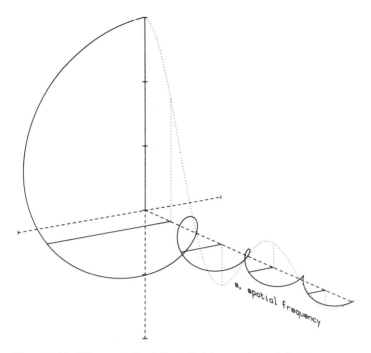

Figure 3.24 The spatially shifted $f(x)$ has a phase shifted spectrum.

3.4.2 Phase Shifts When $2L = 10$, $p = +\frac{1}{2}$

The way to do that would be use the same $+p$, the same width rectangular pulse, but, as we know, to get more spikes we can just increase the period, $2L$, to 10, like the spectrum in Figure 3.16, here repeated as Figure 3.25. Here the period is $2L = 10$, $s_1 = \frac{1}{10}$, the pulse width d is still 1, and there are many more spikes. The general spectral term is, (3.1)

$$\underbrace{\left[\frac{2A}{n\pi}\sin\left(\frac{n\pi d}{2L}\right)\right]}_{a_n}\cos(2\pi s_1 x)$$

The period, $2L$, is 10; $s = \frac{1}{10}$, $d = 1$, so with $A = 1$ the first term will be

$$\underbrace{\left[\frac{2\cdot 1}{1\pi}\sin\left(\frac{1\pi 1}{10}\right)\right]}_{a_n}\cos\left(2\pi \frac{1}{10}x\right) = [0.197]\cos(2\pi(\tfrac{1}{10})x)$$

This spike will be at spatial frequency $s_1 = \frac{1}{10}$, with an amplitude of 0.197. You can see this in Figure 3.25.

78 FOURIER SERIES AND SPECTRA FOR FUNCTIONS OF INFINITE PERIOD

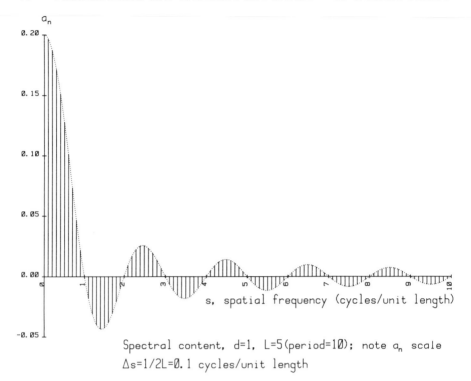

Spectral content, d=1, L=5 (period=10); note a_n scale
$\Delta s = 1/2L = 0.1$ cycles/unit length

Figure 3.25 The longer period pulse has a smaller frequency spacing between spectrum terms.

We will let p again equal $+\frac{1}{2}$; and $d = 1$. The phase shift of the first term at $s = \frac{1}{10}$, will be, as before, $2\pi s_1 \, p$

$$2\pi s_1 \, p = 2\pi(\tfrac{1}{10})(\tfrac{1}{2}) = \pi/10 = 18°$$

The phase shift of the term at $s = \frac{5}{10} = \frac{1}{2}$. will be 90°;

$$2\pi s_5 p = 2\pi(\tfrac{1}{2})(\tfrac{1}{2}) = \pi/2 = 90°$$

The general phase shift is

$$2\pi(n/10)(\tfrac{1}{2}) = n\pi/10, \qquad n = 1, 2, 3, \ldots$$

Let's try it; see Figure 3.26.

We can see the phase shift for each spectral term and an interesting feature: for this particular $f(x)$ and particular $+p$ shift, we see that the first ten terms are shifted from 0° to 180°, in consecutive steps of 18°. The second ten terms, with negative amplitude before the $+p$ shift, have been phase shifted 180° to

PHASE IN INFINITE FOURIER SERIES 79

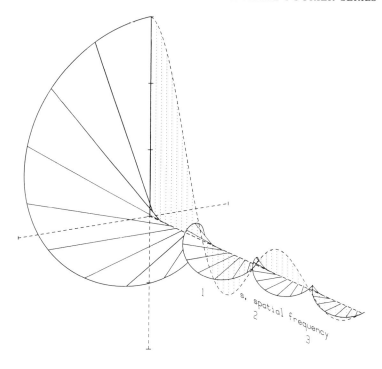

Figure 3.26 The spatially shifted longer period pulse has a phase shifted spectrum.

360°. This particular $f(x)$ and $+p$ cause no spectral spikes to point into the back plane of the diagram. All spikes are perpendicular to the s-axis, though it may not appear so.

One should be able to reason then, that (with the pulse width = 1, shift $p = +\frac{1}{2}$) if we let $2L$ continue to get larger we will simply increase the number of spikes on the "phase surface," but the surface will keep its shape, and no spikes will orient into the back plane under these conditions. Instead of doing that let's increase the shift.

3.4.3 Phase Shifts When $2L = 10$, $p = +1$

If we double the physical shift, make $p = +1.0$, and change nothing else, we will require twice the phase shift. That is, each sinusoidal component of $f(x)$ needs to be shifted twice as much in phase along the x-axis. On our phase surface, the spike rotation will be twice as fast, steps of 36° instead of 18°. Here it is: Figure 3.27. You can see that with $p = +1$, 180° of rotation occurs at $s = \frac{1}{2}$ rather than at $s = 1$ (the first zero) when p was $+\frac{1}{2}$. There is a directly proportional relationship between the amount of physical shift of $f(x)$ and the rate of rotation, and hence the amount of rotation for each spike of the phase

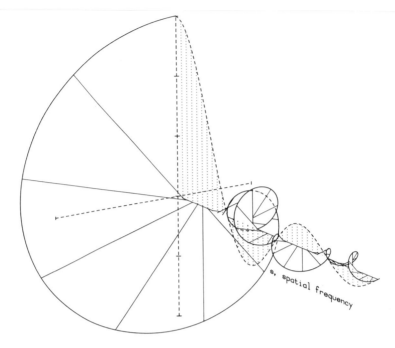

Figure 3.27 Doubling the spatial shift doubles the phase shift of each spectral component.

surface. Twice the physical shift of $f(x)$ means twice the phase shift for each component. This was not too easy to draw; a corkscrew of $\text{sinc}(\pi s)$ amplitude. Careful examination of the surface now shows that some spikes are oriented into the back plane.

3.4.4 Phase Shifts When $p = +\frac{1}{2}$, $2L$ Increasing

Let us return for a moment to the previous smaller shift $p = +\frac{1}{2}$, where the phase shift is easier to see. We have let the period grow to $2L = 40$, hence $s = \frac{1}{40}$, but $d = 1$ so the envelope stays the same. The period being larger decreases the Δs spacing of the spikes to $\frac{1}{40}$ [cycles/unit length]. See Figures 3.28 and 3.29.

Figure 3.30 shows the phase shifted spectrum for the $p = +\frac{1}{2}$ physically shifted $f(x)$. We have provided the non-shifted envelope for comparison (the dashed curve).

These phase shifted spectra are like helixes, (like Figure 3.31) except that the helix has constant amplitude radius. Our spectra have, in general, a changing amplitude, such as $\text{sinc}(\pi s)$.

In Figures 3.32 and 3.33 we have kept the same $f(x)$ and $p = +\frac{1}{2}$, but have increased the period $2L$ to 60, making $s = \frac{1}{60}$, and Δs between spikes

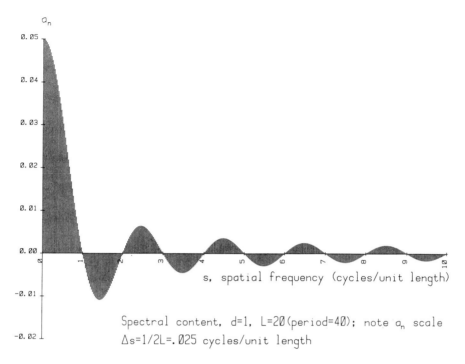

Spectral content, d=1, L=20 (period=40); note a_n scale
$\Delta s = 1/2L = .025$ cycles/unit length

Figure 3.28 The spectrum of a longer period pulse, $f(x)$ without spatial shift.

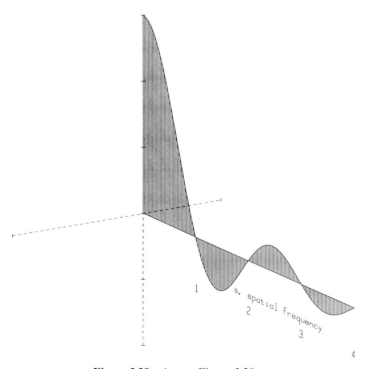

Figure 3.29 As per Figure 3.28.

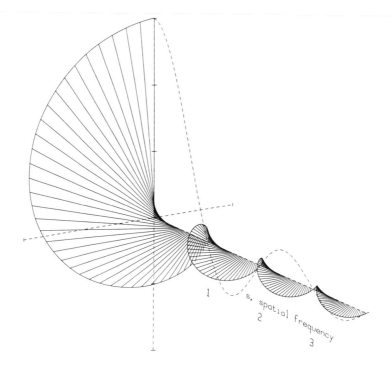

Figure 3.30 $f(x)$, spatially shifted, has a phase shifted spectrum.

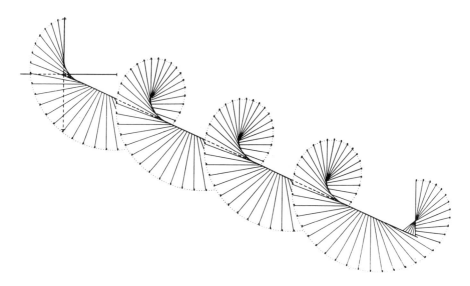

Figure 3.31 The previous figures were based upon first creating a helix plot. These were generally done by "brute force" methods, literally telling the plotter pen where to go, when to lift, etc.

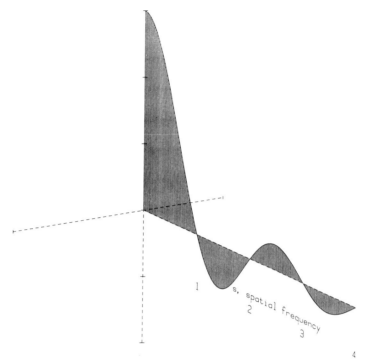

Figure 3.32 As the pulse period increases, the frequency spacing of spectral terms decreases, almost becoming a continuous function. For the spatially shifted $f(x)$ the phase shifted spectrum can be thought of as lying on this screwy warped surface (Figure 3.33).

$\frac{1}{60}$ [cycles/unit length]. If we let the period continue to increase, the spikes become closer yet: with Δs approaching zero, the phase surface becoming continuous as $2L$, the period, becomes infinitely large. This means there is only one single pulse in the space domain of $f(x)$, requiring an infinite number of terms in its spectral representation, $F(s)$; the frequency difference between terms being infinitely small.

3.4.5 Phase Display for Fourier Series

What we have accomplished here is a display on a two-dimensional medium, in three-space, the amplitude spectrum of a one-dimensional infinite Fourier series (which required two-space) and the phase of each component (which required the other space). Very shortly we will be constructing two-dimensional infinite Fourier series (Fourier transforms). Their amplitude display alone will require three-space. The phase aspect will require more space, but we can only draw in three. Hence phase will need to be shown in a separate diagram.

84 FOURIER SERIES AND SPECTRA FOR FUNCTIONS OF INFINITE PERIOD

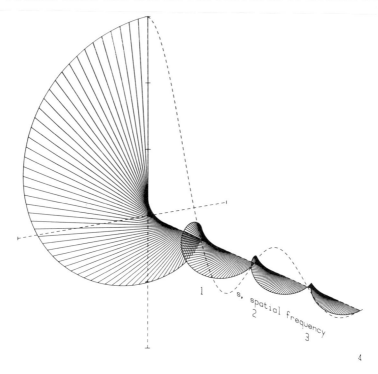

Figure 3.33 As per Figure 3.32.

If the rectangular wave was, say, a sound wave, the physical shift would be unnoticeable to the listener (except during the shift). But certainly the shifted and unshifted waves are different. The only way one could detect the shift would be if one had a "phase detector." We will find this to be true also for our upcoming two-dimensional examples. In the optical diffraction case it seems impossible that the shift is not obviously detectable; it is not obviously detectable without an "optical phase detector."

4

FOURIER SPECTRA FOR NON-PERIODIC FUNCTIONS; ONE DIMENSIONAL

4.1 THE FOURIER TRANSFORM PAIR FOR FUNCTIONS OF INFINITE PERIOD

If the period of the wave is extremely long, sufficiently so that it could be taken as infinitely long, then the wave would no longer be considered to be periodic; the wave becomes non-periodic, a one-shot function, and the spectrum of such a wave becomes a continuous distribution of terms. The spectrum envelope (the shape of the distribution) can often be written in simple form. Figure 4.1 is such a wave, $f(x)$, consisting of only one cycle (the next cycle begins after an infinity of space). It is non-periodic and the equation for the shape of its spectral distribution is, as we will show, the simple expression $(A/k)\sin(dk/2)$, where $k = 2\pi s$. Its shape is plotted in Figure 4.2 as $F(s)$ with $d = 1$.

For the non-periodic wave in the space domain the spectrum has become a continuous distribution of an infinity of wave components. The summation of discrete terms has now become an infinite sum of continuous terms. In general, regardless of the $f(x)$ wave shape, the Fourier series summation which yields the *periodic function*

$$f(x) = a_0/2 + \sum_{n=1}^{n=\infty} [a_n \cos(n\pi x/L) + b_n \sin(n\pi x/L)] \qquad (4.1)$$

(note: this is a repeat of formula (2.1)) becomes the following integral for the

86   FOURIER SPECTRA FOR NON-PERIODIC FUNCTIONS

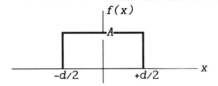

Figure 4.1 This one-shot rectangular pulse never repeats (in our life-space). It has infinite spatial period.

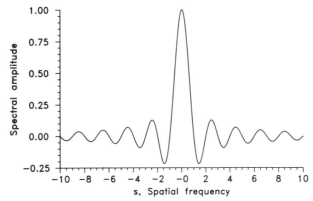

Figure 4.2 The (continuous) spectrum of the rectangular pulse of Figure 4.1.

non-periodic function

$$f(x) = \int_0^\infty [A(k) \cos(kx) + B(k) \sin(kx)] \, dk \qquad (4.2)$$

$A(k)$ and $B(k)$ are the spectrum functions of k, the angular spatial frequency; $k = 2\pi s$.

The discrete summation for the periodic $f(x)$, (4.1), is over all spectral components n. The integration for the non-periodic $f(x)$ is throughout the continuum of angular spatial frequency k. With period essentially infinite it is easy to understand the disappearance of $a_0/2$, the average value for a finite period $f(x)$. Try it. What is the average value of this $f(x)$ (or any one-shot function) over $-\infty$ to $+\infty$? Here is how you get the $A(k)$ and $B(k)$ from $f(x)$:

$$A(k) = (1/\pi) \int_{-\infty}^{+\infty} f(x) \cos(kx) \, dx \qquad (4.3)$$

$$B(k) = (1/\pi) \int_{-\infty}^{+\infty} f(x) \sin(kx) \, dx \qquad (4.4)$$

If the non-periodic function can be defined as an *even function*, these operations simplify; $B(k)$ equals zero. The transformation becomes a pair of Fourier cosine transforms:

$$A(k) = F_c(k) = \sqrt{2/\pi} \int_0^\infty f(x) \cos(kx)\, dx \tag{4.5}$$

$$f(x) = \sqrt{2/\pi} \int_0^\infty F_c(k) \cos(kx)\, dk \tag{4.6}$$

$F_c(k)$ is sometimes called the Fourier cosine transform. Note the symmetry.

If $f(x)$ can be defined as an *odd function*, $A(k)$ equals zero, and we have a pair of Fourier sine transforms:

$$B(k) = F_s(k) = \sqrt{2/\pi} \int_0^\infty f(x) \sin(kx)\, dx \tag{4.7}$$

$$f(x) = \sqrt{2/\pi} \int_0^\infty F_s(k) \sin(kx)\, dk \tag{4.8}$$

$F_s(k)$ is sometimes called the Fourier sine transform. Note the symmetry. If $f(x)$ is even or odd in the space domain, the sine or cosine transforms are straightforward tools with which to obtain the spectrum of $f(x)$. If $f(x)$ is neither odd nor even then both $A(k)$ and $B(k)$ must be evaluated and used to reproduce $f(x)$ using (4.2).

The fact that, in general, both $A(k)$ and $B(k)$ must be used suggests that the Euler relationship could be used to advantage here. When this is done there results a more compact expression for both the spectrum of $f(x)$, $F(k)$, and for $f(x)$ itself:

$$f(x) = 1/\sqrt{2\pi} \int_{-\infty}^{+\infty} F(k) e^{ikx}\, dk \quad \left\{\begin{array}{l}\text{"inverse"}\\ \text{Fourier transform}\end{array}\right\} \tag{4.9}$$

$$F(k) = 1/\sqrt{2\pi} \int_{-\infty}^{+\infty} f(x) e^{-ikx}\, dx \quad \left\{\begin{array}{l}\text{Fourier transform} =\\ \text{spectrum of } f(x)\end{array}\right\} \tag{4.10}$$

This is the symmetric form. These are called a Fourier transform pair. Again we have used notation similar to that of the *Mathematical Handbook of Formulas and Tables* by Murray Spiegel. Other authors may use different, but equivalent, and sometimes nonsymmetric expressions.

This discussion has been carried out in the space–angular-frequency domain (x–k domain) but the reader may have some physical feeling for signals in time (electronic, audio, musical signals, etc.) and may be able to easily visualize or hear harmonic spectra. And indeed in Chapter 1 we chose samples of time signals from common use. Similar Fourier relationships exist for $f(t)$ and $F(\omega)$. Just substitute t for x, s for f, and ω for k; $\omega = 2\pi f$.

The time domain is one dimensional; $f(t)$ and $F(\omega)$ can be displayed on paper of two dimensions, 2-space, where $f(t)$ is the wave in the time domain as a function of t; F, the spectrum, as a function of angular temporal frequency, ω, or temporal frequency, f. The space domain is three dimensional implying that if $f(x, y, z)$ and $F(k_x, k_y, k_z)$ were really functions of 3-space then for their illustration we would really require model building in 4-space, for example: $\{x, y, z, f(x, y, z)\}$. This could become a bit difficult. But Fourier analysis of three-dimensional structures is done, e.g., x-ray diffraction analysis of complex molecules such as DNA. Should a problem require it, there is no reason why Fourier analysis cannot be carried out in 4-, 5-, 6-, or n-dimensional space, at least conceptually. We will limit ourselves here to the Fourier analysis of no more than two-dimensional structures. The references contain details of books dealing with the analysis of three-dimensional forms, i.e., x-ray diffraction.

The remainder of this book deals with Fourier analysis mainly in the space–spatial-frequency domain.

There are at least two other equivalent ways of writing Fourier transform pairs in 2-space:

$$f(x) = \int_{-\infty}^{+\infty} F(k)e^{ikx}\, dk \qquad \left\{\begin{array}{c}\text{"inverse"}\\ \text{Fourier transform}\end{array}\right\} \quad (4.11)$$

$$F(k) = (1/2\pi) \int_{-\infty}^{+\infty} f(x)e^{-ikx}\, dx \qquad \{\text{Fourier transform}\} \quad (4.12)$$

This is the nonsymmetric form. The product of the factors before the integral sign must be $(\frac{1}{2}\pi)$. $(1/\sqrt{2\pi})$ before each is the symmetric form. It is also arbitrary as to which integral contains the negative exponential. But whichever, one must be consistent throughout usage.

Here is another symmetric Fourier transform pair using the spatial frequency s rather than the angular spatial frequency k. In this form the 2π factors are in the exponents rather than before the integral sign.

$$f(x) = \int_{-\infty}^{+\infty} F(s)e^{i2\pi sx}\, ds \qquad \left\{\begin{array}{c}\text{"inverse"}\\ \text{Fourier transform}\end{array}\right\} \quad (4.13)$$

$$F(s) = \int_{-\infty}^{+\infty} f(x)e^{-i2\pi sx}\, dx \qquad \{\text{Fourier transform}\} \quad (4.14)$$

One needs to use caution when reading different authors; other variables are also often used instead of k, x, s; e.g., η, ξ, α, f_x, etc. And as Champeney (1973) has noted, "There is no universally accepted convention governing the definitions of the terms 'Fourier transform' and 'inverse transform'." A double-headed arrow could be drawn between such a transform pair, for just as $F(k)$ is the Fourier transform of $f(x)$ and $f(x)$ is the "inverse" Fourier transform of $F(k)$, it is also true that if $f(x)$ is the Fourier transform of $F(k)$ then $F(k)$ is the "inverse" Fourier transform of $f(x)$. It is simply a matter of how one defines $f(x)$ and $F(k)$. The constant 2π always shows up, somewhere, in these strictly mathematical formulations of Fourier transform pairs.

There being no universally accepted definition for a Fourier transform or an inverse transform allows us some options regarding the form and placement of the quantity 2π, and which of the transform pair is to be called the inverse transform, the one with the positive or the one with the negative exponential. Certainly, if one intends to move back and forth between a transform pair, then one must choose a specific acceptable form and stick with it; that will establish specific placement of the 2π constant(s), and which one of the transform pair has the negative sign in the exponential. But if one desires to work with only one of the pair, then it would not matter which form was used, any would be legitimate. You could choose not to use a constant, 2π, in a nonsymmetric form and still have freedom to use either $\pm$ exponential.

Clearly then, what is important in a Fourier transform is the integral itself, not the constant. And the fact that either a $\pm$ exponential can be chosen strongly suggests that the square of the transform may be more consistently meaningful than the transform itself.

Suppose you did a mathematical solution to a physical problem, perhaps starting with Newton's laws or Maxwell's equations, and the result you get has this form,

$$\text{Result} = (\text{constant}) \cdot \int_c^{c+g} q(x)e^{i\alpha x}\,dx$$

Would you be justified in claiming that a Fourier transform relationship exists between $q(x)$ and your result? I hope so, for that is the claim we will make in Chapter 5.

4.2 THE FOURIER TRANSFORM AND SPECTRUM OF A SQUARE PULSE

We begin this section with a specific example of calculation and display. Consider the non-periodic function shown in Figure 4.3 (all writers seem to use this example). Note the change in notation; the d pulse width of Chapter 3 and Figure 4.1 is now L. We will calculate the Fourier transform (the spectrum) of this space domain function by two methods. First using the Fourier cosine transform (the sine transform will be zero for this even function), then by using the complex Fourier transform—the symmetric form with negative exponential.

90 FOURIER SPECTRA FOR NON-PERIODIC FUNCTIONS

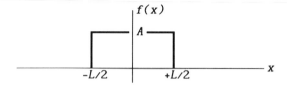

Figure 4.3 We have redefined the pulse width of Figure 4.1 from d to L.

1. Using the cosine transform:

$$F(k) = F_c(k) = \sqrt{2/\pi} \int_0^{+\infty} f(x) \cos(kx)\, dx = \sqrt{2/\pi} \int_0^{+L/2} A \cos(kx)\, dx$$

$$= \sqrt{2/\pi}(A/k) \sin(kx) \Big|_0^{+L/2} = \sqrt{2/\pi}(A/k)[\sin(kL/2) - 0]$$

$$F(k) = \sqrt{2/\pi}(A/k) \sin(kL/2) \cdot (L/2)/(L/2) = (AL/\sqrt{2\pi}) \frac{\sin(kL/2)}{(kL/2)}$$

$$\boxed{F(k) = (AL/\sqrt{2\pi}) \operatorname{sinc}(Lk/2)} \qquad (4.15)$$

This is plotted in Figure 4.4.

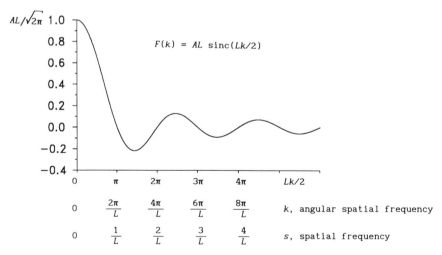

Figure 4.4 The Fourier transform (only positive frequencies) of the $f(x)$ spatial pulse of Figure 4.3. Notice that the first zero value occurs at a spatial frequency, s, equal to $1/L$, for this specific pulse.

THE FOURIER TRANSFORM AND SPECTRUM OF A SQUARE PULSE

2. Using the exponential transform:

$$F(k) = (1/\sqrt{2\pi}) \int_{-\infty}^{+\infty} f(x) e^{-ikx} \, dx = (\sqrt{1/2\pi}) \int_{-L/2}^{+L/2} A e^{-ikx} \, dx$$

$$= (-A/ik\sqrt{2\pi}) e^{-ikx} \Big|_{-L/2}^{+L/2} = (-A/ik\sqrt{2\pi})(e^{-ikL/2} - e^{+ikL/2})$$

$$= (2A/k\sqrt{2\pi}) \left(\frac{e^{+ikL/2} - e^{-ikL/2}}{2i} \right) = (2A/k\sqrt{2\pi}) \sin(kL/2) \cdot (L/2)/(L/2)$$

$$F(k) = (AL/\sqrt{2\pi}) \frac{\sin(kL/2)}{(kL/2)}$$

$$\boxed{F(k) = (AL/\sqrt{2\pi}) \operatorname{sinc}(Lk/2)} \qquad (4.15)$$

Had this been a function of time, $f(t)$, T wide, the result would have been:

$$F(\omega) = (AT/\sqrt{2\pi}) \operatorname{sinc}(T\omega/2) \qquad (4.16)$$

The spectra would be displayed as in Figure 4.5. The spectrum is a continuum because the period has become infinite; the frequency spacing between spectral spikes, Δs, has become zero. Figures 4.4 and 4.5 show only the positive frequencies.

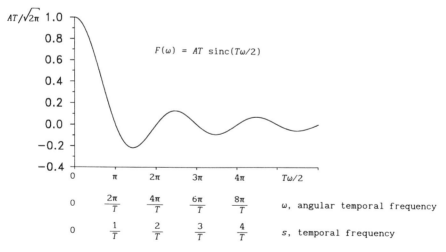

Figure 4.5 Had the pulse of Figure 4.3 occurred in the time domain, an $f(t)$ of width T, its spectrum of temporal frequencies (only the positive ones) would appear as shown here. Notice that the first zero value occurs at the very simply expressed temporal frequency, f, equal to $1/T$, for this specific pulse.

92 FOURIER SPECTRA FOR NON-PERIODIC FUNCTIONS

Pay particular attention to the simple numerical horizontal scale made up by using temporal frequencies, f, and spatial frequencies, s. Figure 5.13 will illustrate conversion of the frequency axis to other scales.

Keep in mind the following mathematical form which we now know yields the spectrum of the function $f(x)$ and is called the Fourier transform of $f(x)$:

$$F(k) = (1/\sqrt{2\pi}) \int_{-\infty}^{+\infty} f(x) e^{-ikx} \, dx \qquad (4.17)$$

In much of the work that follows we will be able to use the nonsymmetric form and drop the constant before the integral.

4.3 PHASE IN ONE-DIMENSIONAL FOURIER TRANSFORMS

We now ask an interesting question. Having found the Fourier transform for the rectangular pulse of width L, and A high, do you suppose the spectral content will be any different for the same pulse shifted $+p$ along the x-axis? See Figure 4.6. Surely the same waves would be required to create both. But what is different? Here is the calculation for the $+p$-shifted $f(x)$:

$$F(k) = (1/\sqrt{2\pi}) \int_{-\infty}^{+\infty} f(x) e^{-ikx} \, dx = (1/\sqrt{2\pi}) \int_{-L/2+p}^{+L/2+p} A e^{-ikx} \, dx$$

$$= (1/\sqrt{2\pi}) \left(\frac{-A}{ik}\right) [e^{-ik(L/2+p)} - e^{-ik(-L/2+p)}]$$

$$= (1/\sqrt{2\pi}) \left(\frac{+A}{ik}\right) e^{-ikp} [e^{ikL/2} - e^{-ikL/2}]$$

Multiplying by $\frac{2}{2}$,

$$F(k) = (1/\sqrt{2\pi}) \left(\frac{2A}{k}\right) e^{-ikp} \left[\frac{e^{ikL/2} - e^{-ikL/2}}{2i}\right]$$

$$= (2A/k\sqrt{2\pi}) e^{-ikp} \cdot \sin(kL/2) = (AL/\sqrt{2\pi}) e^{-ikp} \cdot \frac{\sin(kL/2)}{kL/2}$$

$$F(k) = e^{-ikp} (AL/\sqrt{2\pi}) \operatorname{sinc}(kL/2) \qquad (4.18)$$

The spectral content is identical to that of the *symmetrical* real function $f(x)$, but now, for the $+p$-shifted function, no longer symmetrical, there is an additional factor, called a phase factor, e^{-ikp}.

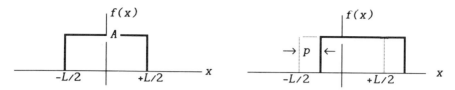

Figure 4.6 The symmetric rectangular pulse, $f(x)$, and then shown spatially shifted $+p$. This will introduce a phase shift factor into the Fourier transform of the shifted pulse.

[**Caution:** If we had used the Fourier transform with positive exponential in the integral (e^{+ikx}) the phase factor would have been e^{+ikp}. Just one more instance to point out how one must be consistent with Fourier techniques, there being no international standards of definition.]

If the Fourier transform uses the negative exponential and the $f(x)$ function is $+p$ shifted, i.e., $f(x - p)$, then the phase factor will have a negative exponent,

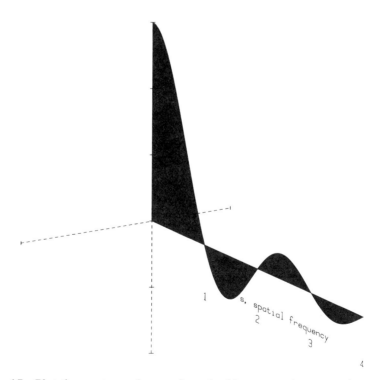

Figure 4.7 Plot the spectrum, the transform, in this manner, in preparation to show the phase of each component. Relaxing the period of the pulse, bringing it from infinity to 90 creates these black blobs. 90 terms occur before the first zero. No phase shown here.

94 FOURIER SPECTRA FOR NON-PERIODIC FUNCTIONS

e^{-ikp}. The phase *function* itself is

$$\phi(k) = +pk \qquad (4.19)$$

For the conditions stated, the phase factor is $e^{-i\phi(k)} = e^{-ipk}$. Since, in general, $k = 2\pi s$

$$\phi(s) = +p2\pi s \qquad (4.20)$$

This gives the phase of each spectral component as a function of s.

With the pulse width $L = 1$ (like $d = 1$ in Chapter 3), and period very large (not infinite but, say 90), the fundamental spectral frequency, $s = \frac{1}{90}$ and $\Delta s = \frac{1}{90}$ [cycles/unit length]. This means that there are 90 terms before sinc(πs) has zero amplitude, see Figure 4.7. If $f(x)$ is $+p$ shifted, $f(x - p)$, and again $p = \frac{1}{2}$, the phase function becomes

$$\phi(s) = +2\pi ps = +2\pi(\tfrac{1}{2})s = \pi s$$

For the spatial frequency component at $s = 1$, the phase shift (rotation) has become π, or 180°, as before. Figure 4.8 shows these results.

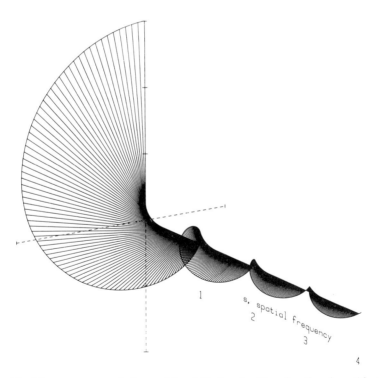

Figure 4.8 The spectrum and phase of the shifted pulse. Showing the phase (the twist) eliminated the black blobs of Figure 4.7. The pulse period has been relaxed bringing it from infinity to 90.

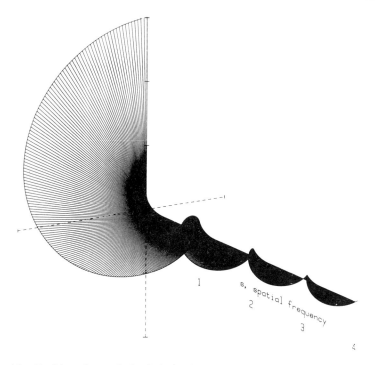

Figure 4.9 Pushing the period of the pulse back out to 180 allows us to see the approaching continuous surface, if we let the period go out to infinity. Let your eye fill in the gaps.

Figure 4.9 shows the same $f(x)$ and $+p$ shift $= \frac{1}{2}$, with a period equal to an even larger value, 180. The frequency spacing is $\frac{1}{180}$ [cycles/unit length]. 180 terms occur before reaching the component at $s = 1$ where $\phi(s) = \pi$ or 180°.

For infinite period, one can imagine the spectrum, with or without shift, becoming a continuous surface, a continuous function, as Δs becomes zero.

For this one-dimensional function, $f(x)$, the phase function, $\phi(s)$, can be displayed as in Figure 4.10.

$$\phi(k) = +pk \qquad \text{or} \qquad \phi(s) = +p2\pi s$$

With $p = +\frac{1}{2}$ and a period of 180,

$$\phi(s) = (+\tfrac{1}{2})2\pi s = \pi s = 180° \cdot s$$

For the spatial frequency component at $s = 1$, its phase shift is π or 180°.

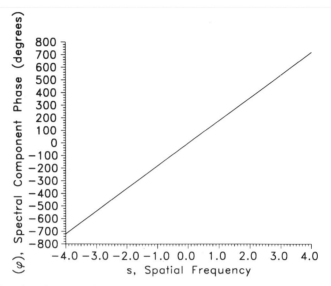

Figure 4.10 The phase continuum can also be put into a separate plot. The linear spatial shift gives rise to a linear phase shift, and we can even include the negative frequencies.

4.4 THE REAL AND IMAGINARY PARTS OF FOURIER SPECTRA AND SERIES

If all the spectral components of these one-dimensional Fourier transforms were of the same phase, then there would be no need for the twisting helical shape of our spectral diagrams which illustrated both amplitude and phase of our spectral components. In Section 3.4, when sinusoidal functions were used for Fourier series, the presence of phase for each spectral wave component showed up in each term, such as

$$\cos[2\pi n s_1 (x - p)] = \cos(2\pi n s_1 x - 2\pi n s_1 p)$$

The $(2\pi n s_1 p)$ represents the phase of the nth component. The quantities p, the shift, and s_1, the fundamental frequency, are fixed, so the phase increases with each term as n increases at each term.

When using exponential notation the phase of each component shows up as a complex exponential, as in the example of the last section, where the phase factor was e^{-ikp}. This can be translated into sinusoidal form through the Euler formula to give $\cos(kp) - i\sin(kp)$. Whether the i will disappear or not depends upon the original nature of $f(x)$. When a computer is used to do Fourier analysis it usually creates its answers in complex (exponential) form. This means that for each value of k (or s) there could be both a real part and an imaginary part. Computer software will tabulate both, and will compose diagrams of either

THE REAL AND IMAGINARY PARTS OF FOURIER SPECTRA AND SERIES   97

the real part or the imaginary part, or the square root of the sum of the squares of these two parts. Some software can calculate the phase at each k, phase = arctan[imaginary part/real part], and plot that diagram, phase versus k.

For the unshifted rectangular pulse of the last section, we had in the exponential calculation of the transform

$$(-A/ik\sqrt{2\pi})(e^{-ikL/2} - e^{+ikL/2})$$

Even with the presence of all those i's this function is completely real. Let's multiply by i/i, and "Eul" the exponentials (i.e., use the Euler identity):

$$(-A/ik\sqrt{2\pi})\left(\frac{i}{i}\right)[\cos(kL/2) - i\sin(kL/2) - \cos(kL/2) - i\sin(kL/2)]$$

$$= (+Ai/k\sqrt{2\pi})[-2i\sin(kL/2)] = (+2A/k\sqrt{2\pi})\sin(kL/2)$$

This is completely real. A and L being given quantities, a tabulation of values of this transform as a function of k will contain only real parts.

On the other hand, the calculation of the transform for the $+p$-shifted $f(x)$ led to

$$(1/\sqrt{2\pi})\left(\frac{+A}{ik}\right)e^{-ikp}[e^{ikL/2} - e^{-ikL/2}]$$

We will "Eul" this one twice; first,

$$(1/\sqrt{2\pi})\left(\frac{+A}{ik}\right)e^{-ikp}[\cos(kL/2) + i\sin(kL/2) - \cos(kL/2) + i\sin(kL/2)]$$

$$= (1/\sqrt{2\pi})\left(\frac{+A}{ik}\right)e^{-ikp}[2i\sin(kL/2)] = (2/\sqrt{2\pi})\left(\frac{A\sin(kL/2)}{k}\right)e^{-ikp}$$

Now "Eul" the second time.

$$(1/\sqrt{2\pi})\left(\frac{2A\sin(kL/2)}{k}\right)[\cos(kp) - i\sin(kp)]$$

$$= \frac{2A\sin(kL/2)\cos(kp)}{k\sqrt{2\pi}} - i\left(\frac{2A\sin(kL/2)\sin(kp)}{k\sqrt{2\pi}}\right)$$

The transform for the $+p$-shifted rectangular $f(x)$ for all values of k consists of a real and an imaginary part. The phase of the component at each k can be obtained from the arctan formula. Computer software will usually tabulate both the real and imaginary parts.

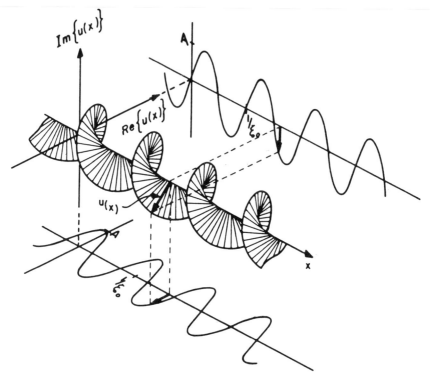

Figure 4.11 Gaskill's depiction of $Ae^{+i2\pi\xi_0 x}$. Of course, hidden in this exponential are the two separate parts, the real part and the imaginary part. (J. Gaskill, *Linear Systems, Fourier Transforms, and Optics*, copyright © 1978, reprinted by permission of the publisher, John Wiley and Sons, Inc.)

Though not the first to do so, Gaskill (1978) has shown how to plot complex functions, using the helix idea and two additional sets of axes. Figure 4.11 is his depiction of

$$u(x) = Ae^{+i2\pi\xi_0 x} = A[\cos(2\pi\xi_0 x) + i\sin(2\pi\xi_0 x)]$$

That's pretty good artwork, I believe done by hand by Don Cowen.

5

THE DIFFRACTION OF LIGHT AND FOURIER TRANSFORMS IN TWO DIMENSIONS

In this chapter we will enter the realm of two-dimensional non-periodic functions, $f(x, y)$, and learn how to calculate their Fourier transforms. We will show that, under proper conditions, when light passes through a two-dimensional aperture and is scattered by the aperture structure (diffracted), the intensity in the visible Fraunhofer diffraction pattern is directly proportional to the square of the Fourier transform of the two-dimensional aperture function, which we will call $A(x, y)$. Thus the Fraunhofer diffraction pattern will contain information about the spatial frequency spectrum of $A(x, y)$. It will be necessary to introduce some new notation for this new dimensionality.

5.1 THE DIFFRACTION OF LIGHT

Plane waves (waves of planar constant amplitude) of wavelength λ and propagation constant or angular spatial frequency k (equal to $2\pi/\lambda$) are incident normally (perpendicularly) on a screen Σ at the x, y plane with an aperture A in it (Figure 5.1). The *amplitude* per unit area of the optical disturbance emerging from the aperture is $\mathscr{E}(x, y)$. The units of $\mathscr{E}(x, y)$ are [amplitude/m^2]. (Section 5.7 is a discussion concerning the "amplitude" of a light wave.) We will allow this amplitude to vary on emergence from the aperture due to aperture structure, i.e., $\mathscr{E}(x, y)$ (originally constant over the incident planar wavefronts) becomes $\mathscr{E}(x, y)$, dependent on aperture coordinates x, y, upon emergence from the aperture. Later we will allow the aperture phase structure to be dependent on x, y also; we will use aperture phase delays.

Thus, with light waves of amplitude per unit area $\mathscr{E}(x, y)$ emerging from the

THE DIFFRACTION OF LIGHT AND FOURIER TRANSFORMS

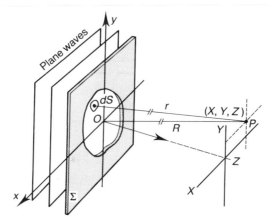

Figure 5.1 Plane waves of light incident on an aperture are diffracted, or scattered, by the aperture and the result will be examined some distance farther along the optical Z-axis in the X, Y plane. The coordinates in the aperture plane are x, y.

aperture, our problem is to determine the wave *amplitude* further along the Z-direction in the X, Y plane. We use capital letters in the more distant plane. We will eventually limit ourselves to points $P(X, Y)$ such that $Z \cong R$, and $R^2 \gg (x^2 + y^2)$. The variables x and y are coordinates within the aperture and R is the distance from O to P. We will provide reasons for the approximations shortly.

Consider infinitely small differential areas of the aperture dS (essentially point sources emitting what are often termed spherical "Huygens'" wavelets), where the emergent wave (disturbance) amplitude is $\mathscr{E}(x, y)\, dS$, with units of [(amplitude/m²)·m²]. We let these spherical Huygens' wavelets emanating from all dS of the entire aperture propagate out to the point $P(X, Y)$ where the net effect is obtained by the principle of superposition, i.e., by summing the *amplitudes* from all the spherical wavelets from the entire aperture. This reshaping of the light beam as it passes through such an aperture (without phase structure) is called diffraction. "Diffraction" is derived from the Latin and means "breaking up." At the aperture, *conceptually*, the originally planar light wave front is broken up into infinitely small spherical wavelets, Huygens' wavelets, the concept of Figure 5.2. We consider only that part of the wavelets propagating within small angles of the optical Z-axis, i.e., propagation mainly in the forward direction. The amplitude of the spherical Huygens' wavelets must decrease as $1/r$, (see below) so *from each element dS of the aperture* there will be received at point $P(X, Y)$ a wave contribution which we shall call the differential dE:

$$dE = \frac{\mathscr{E}(x, y)\, dS}{r} \cos(\omega t - kr)$$

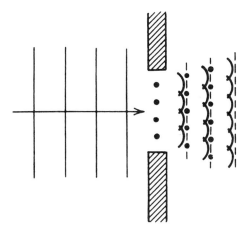

Figure 5.2 The concept of Huygens' wavelets is employed; these will need to be summed farther along in Z at points in the X, Y plane. (W. L. Stutzman and G. A. Thiele, *Antenna Theory and Design*, copyright © 1981, reprinted by permission of the publisher, John Wiley and Sons, Inc.)

dE has units of [(amplitude/m² · m) · m²], i.e., [amplitude/m]. We have used E because the amplitude of light is often taken as a measure of the electric field of the associated light wave. We rewrite this in rearranged exponential form,

$$dE = \frac{\mathscr{E}(x, y)}{r} \operatorname{Re}\{e^{i(\omega t - kr)}\} \, dS$$

We have just used the Euler identity

$$e^{i\theta} = \cos\theta + i\sin\theta, \qquad \text{with } \theta = (\omega t - kr)$$

"Re" means "the real part of." We only want the cosine part; $\sin\theta$, without the i, is called the imaginary part. Breaking up the exponential we have

$$dE = \frac{\mathscr{E}(x, y)}{r} \operatorname{Re}\{e^{i\omega t} \cdot e^{-ikr}\} \, dS \tag{5.1}$$

We will eventually be able to refer all wavelets to a time, $t = 0$. That will help since $e^0 = 1$. At point $P(X, Y)$ the differential amplitude element, dE, from the *spherical* wavelets must go as $1/r$ because the square of dE is related to the radiated power in Watts per square meter, which we know varies as $1/r^2$ for spherical waves from point sources.

We will need a value for r expressed in terms of geometrical coordinates. If dS is located at (x, y), R is the distance from point O to point P, and r is the distance from dS to P, in Figure 5.1, then

$$R^2 = X^2 + Y^2 + Z^2$$
$$r^2 = (X - x)^2 + (Y - y)^2 + Z^2$$

Solving for Z^2, we can rewrite r^2,

$$Z^2 = R^2 - X^2 - Y^2$$
$$r^2 = R^2 - X^2 - Y^2 + (X - x)^2 + (Y - y)^2$$

Squaring the indicated quantities and collecting terms,

$$r^2 = R^2 - 2Xx - 2Yy + x^2 + y^2$$

We put this in a more useful form,

$$r^2 = R^2\left[1 - \frac{2Xx + 2Yy}{R^2} + \frac{x^2 + y^2}{R^2}\right]$$

We take the square root to obtain the exact value for r,

$$r = R\left[1 - \frac{2Xx + 2Yy}{R^2} + \frac{x^2 + y^2}{R^2}\right]^{1/2} \quad \text{exactly}$$

This is an exact expression for r. We insert this into (5.1), treat Re as "understood," and use $dx\,dy$ for dS, to obtain dE at $P(X, Y)$

$$dE = \frac{\mathscr{E}(x, y) \cdot e^{i\omega t} \cdot e^{-ikR[1 - 2(Xx + Yy)/R^2 + (x^2 + y^2)/R^2]^{1/2}}}{R[1 - (2Xx + 2Yy)/R^2 + (x^2 + y^2)/R^2]^{1/2}} dx\,dy \quad (5.2)$$

dE is here expressed in terms of the integration variables, x and y. It can be seen that it might be a bit difficult to sum all the dE, so let us now make the approximations referred to earlier. The units of dE remain [amplitude/m].

Other writers deriving this via a different approach often generate additional factors, e.g., $(-i)$, $(1/\lambda)$, and an "obliquity" factor. It will become clear that *for our purposes* each of these can be considered constant, but for correct dimensionality the $1/\lambda$ and its m^{-1} dimension must be considered as included in our $\mathscr{E}(x, y)$ quantity. It provides one of the m^{-1} of the [amplitude/m^2].

5.2 THE FRAUNHOFER APPROXIMATION AND TWO-DIMENSIONAL FOURIER TRANSFORMS

The denominator of the dE expression, (5.2), is simply r; this $1/r$ factor is a measure of the *decrease in amplitude* with distance for spherical wavelets. If point $P(X, Y)$ is chosen near coaxial and the aperture dimensions in Σ are small compared to R, and R will be constant, then for all wavelets we will set $r = R$ *in this factor*—another constant.

The exponential term, $e^{-ikR[1-\cdots]}$, requires more careful consideration. This quantity contains the r distances for calculating the interference effects of each dE at $P(X, Y)$, so here we would not want our approximation to be inaccurate by more than a very small fraction of λ, the wavelength of the light incident on the aperture. Let us try an approximation and attempt to validate it afterwards. We have the exponential

$$e^{-ikR[1 - 2(Xx+Yy)/R^2 + (x^2+y^2)/R^2]^{1/2}}, \qquad \text{exact} \qquad (5.3)$$

First let us consider the *far field case* where $x^2 + y^2 \ll R^2$ and use this as a justification for considering the third term to be negligible. (Then for our validation test we expect R to be large. How large? We will see.) So now we have

$$e^{-ikR[1 - 2(Xx+Yy)/R^2]^{1/2}} \qquad \{\text{in the far field case}\} \qquad (5.4)$$

We use the binomial expansion of $(1-x)^{1/2} = 1 - x/2 - x^2/8 - \cdots$, and keep only the first two terms, yielding

$$e^{-ikR[1 - 2(Xx+Yy)/2R^2]}$$

Canceling the 2's of the second term,

$$e^{-ikR[1 - (Xx+Yy)/R^2]} \qquad \begin{cases} \text{Far field case and} \\ \text{binomial expansion} \end{cases} \qquad (5.5)$$

Taking R inside the brackets yields

$$e^{-ik[R - (Xx+Yy)/R]}$$

Then with $r = R$, dE, (5.2), becomes

$$dE = \frac{\mathscr{E}(x, y) \cdot e^{i\omega t} \cdot e^{-ik[R-(Xx+Yy)/R]}}{R} \, dx \, dy$$

With a slight rearrangement,

$$dE = \frac{\mathscr{E}(x, y) \cdot e^{i(\omega t - kR)} \cdot e^{+ik(Xx + Yy)/R}}{R} \, dx \, dy \tag{5.6}$$

It will be noted that we have moved the $-kR$ part of the second exponential to the first exponential factor; this is in preparation for treating the first exponential as a constant. The quantity $k(Xx + Yy)/R = (2\pi/\lambda R)(Xx + Yy)$ is dimensionless as it should be.

But now we wish to examine the region of validity for our approximations. How large must R be? We reduced the quantity

$[1 - 2(Xx + Yy)/R^2 + (x^2 + y^2)/R^2]^{1/2}$ {Exact, from (5.3)}

$[1 - 2(Xx + Yy)/R^2]^{1/2}$ {Far field case, (5.4)}

$[1 - (Xx + Yy)/R^2]$ {Far field case and binomial expansion, (5.5)}

These quantities, multiplied by kR, i.e., multiplied by $(2\pi/\lambda)R$, represent the phase of the light waves at points (X, Y) in the plane at a distance R from the aperture plane, x–y. For our work it will be necessary that we use the form (5.5), so, in use, the phase values should not differ from those which would be calculated from the exact expression, (5.3). If we seek agreement to within 0.1%, using light of 6000 Å would mean that our wave phases would be incorrect by only 6 Å; not too bad. Perhaps 0.01% would be more desirable. How large should R be to achieve such agreement? This is not a terribly straightforward problem to solve.

A general statement for the phase error across the pattern of $E(x, y)$ is not easy to write or explain. With the approximations made, the phase of $E(x, y)$ will be correct on a (in general) curved surface. We want to treat the results as correct on a plane surface. In doing so, by how much is the phase in error? Elimination of phase error is mentioned in Section 6.1.3. We can suggest some conditions which will create small phase errors.

We could specify that the aperture sizes, x, and y, not exceed 0.01 meters; and that you agree not to look at diffraction in X–Y at greater than 0.025 meters from the optical Z-axis. One can also choose the wavelength, perhaps $\lambda = 6328$ Å. Weaver (1983) addresses these validity questions, as do Gaskill (1978) and Goodman (1968). The necessary R distances to make the Fraunhofer approximations valid are typically of the order of 1000 meters and more; although one does have the option of varying it by selecting other values of maximum x and y, maximum X and Y, and λ.

1000 meters or more? If we are to do this on an optical bench we will need a very, very long one. But we shall find that there is another way. We will show in Section 5.6 that our approximation will be a good one simply by using a lens of focal length F behind the aperture. The pattern we seek will be present

THE FRAUNHOFER APPROXIMATION   105

in the focal plane of the lens, at a distance F from the lens; easy to put on a conventional optical bench.

These approximations specify regions where the diffraction pattern of light is termed "Fraunhofer" type. Generally, if the approximations are not valid, e.g., closer to the aperture, then the diffraction is of the "Fresnel" type, see Figure 5.3. Chapter 11 contains a brief comment and example of Fresnel diffraction by Fourier methods. Shack and Harvey (1975) describe an approach which is useable everywhere, from infinity right up to the aperture.

Goodman (1968), using a somewhat different approach, provides an equation of the form

$$R \gg \frac{k(x^2 + y^2)_{max}}{2}$$

to specify the condition for Fraunhofer diffraction. The quantities x and y are the maximum halfwidths of the aperture, and $k = 2\pi/\lambda$.

Within the limits of our approximation then, at $P(X, Y)$

$$dE = \frac{\mathscr{E}(x, y)}{R} \cdot e^{i(\omega t - kR)} \cdot e^{ik(Xx + Yy)/R} \, dx \, dy$$

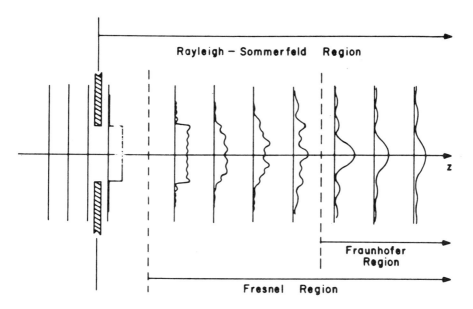

Figure 5.3 Depending on the distance from the aperture, different approximations are made to determine the light distribution in an $X-Y$ plane. We will be mainly concerned with the Fraunhofer region (J. Gaskill, *Linear Systems, Fourier Transforms, and Optics*, copyright © 1978, reprinted by permission of the publisher, John Wiley and Sons, Inc.)

106   THE DIFFRACTION OF LIGHT AND FOURIER TRANSFORMS

is the contribution from each aperture element $dx\,dy$. The total effect at $P(X, Y)$ is a sum of all the dE emitting over the entire aperture.

$$E(\text{total at } P) = \int dE \quad \text{(integrated over the aperture)}$$

$$E(X, Y) = \frac{1}{R} \cdot e^{i(\omega t - kR)} \cdot \iint_{\text{aperture}} \mathscr{E}(x, y) e^{ik(Xx + Yy)/R}\, dx\, dy \quad (5.7)$$

Under our approximations, this integral is correct to within a constant factor. The quantity $e^{i(\omega t - kR)}$ references the wave at $P(X, Y)$ to some fixed time and gives the time variation of the total optical disturbance at $P(X, Y)$. For the present we are not interested in this; we can set our clock to $t = 0$ when we examine $E(X, Y)$. The quantities k and R are constant, thus the factors before the integral are constant. We are interested only in relative values between different $P(X, Y)$ so we set $(1/R)\exp[i(\omega t - kR)] = 1$.

The light waves, as they leave (pass through) the aperture will be altered by the aperture itself. We describe this with an aperture function $A(x, y)$ which allows the amplitudes of $\mathscr{E}(x, y)$ to vary over the aperture. In other words $\mathscr{E}(x, y)$ becomes $A(x, y)$ and as such needs to remain inside the integral (summation). Putting all this together:

$$E(X, Y) = \iint_{\text{aperture}} A(x, y) \cdot e^{ik(Xx + Yy)/R}\, dx\, dy \quad (5.8)$$

Under our approximations, (5.8) correctly provides *relative values* for the total light amplitude at points in the X–Y plane.

We will illustrate in Chapter 9 how a wave may be slowed down (delayed) at the aperture by introducing a phase function into the aperture; such apertures can be described with real and imaginary numbers, i.e., complex algebra, and such apertures are termed complex. For now, $A(x, y)$ will consist of only real numbers and is therefore termed a real aperture.

This last integral should look familiar. Is it a Fourier transform? Yes it is, and two dimensional at that. With the positive exponential some writers refer to it as the nonsymmetric form. A negative exponential would have been proper also. This is what we were after, and why we made all the approximations. If $A(x, y)$ describes an aperture then the Fourier transform of $A(x, y)$ will represent the relative values of the amplitude of the diffracted light, sometimes called the optical disturbance, in a Fourier plane—a plane where the approximations are valid. With the additional factor $e^{i\omega t}$ included, (5.8) would provide relative *instantaneous* amplitude values. The Fourier plane meets the

R approximation conditions previously calculated, 1000 meters or more from the aperture, or in the focal plane of a lens placed after the aperture, as we show in Section 5.6.

In Chapter 4 we discussed the uncertain definition of a Fourier transform pair, with various forms of 2π, and plus or minus exponentials. In the physical diffraction derivation just completed we achieved the same integral form but with a different constant than 2π. Nevertheless, the claim is made that the "Fraunhofer amplitude" of the diffracted light is obtained, within a multiplicative constant, by a Fourier transform of the aperture function, $A(x, y)$. The 2π fits within that constant. And we should then be able to "look upon" the Fraunhofer amplitude as a spatial frequency "spectrum" of the (spatial) aperture function. There is the challenging problem that at optical frequencies the amplitude of light can't be detected; we'll need to examine the consequences of that. But you should feel some confidence, at this point, that if you take $E(X, Y)$ of (5.8) and put it into a Fourier integral *with negative exponential*, and do the inverse transformation, then solution of the integral will produce the aperture function $A(x, y)$ within a multiplicative constant. We will usually follow an established practice and normalize the spectra we create (the diffraction patterns), normalizing them to unity, i.e., 1.0 at their maximum values, usually at the center of the function, at $(X, Y) = (0, 0)$.

5.3 EXPLANATION OF NOTATION TO BE USED

If $A(x, y)$ of (5.8) were separable into $A_x(x) \cdot A_y(y)$ we could write

$$E(X, Y) = \left[\int A_x(x) e^{ikXx/R} \, dx\right] \cdot \left[\int A_y(y) e^{ikYy/R} \, dy\right] \quad (5.9)$$

and we would have the product of two one-dimensional Fourier transforms, making it more simple. We can refer to this possibility as a Separation Theorem. This is perhaps the only situation where this two-dimensional problem can be reduced to two one-dimensional problems.

But our earlier one-dimensional mathematical example, with the square pulse, L wide, in Section 4.2, was written, (4.15),

$$F(k) = (1/\sqrt{2\pi}) \int_{-\infty}^{+\infty} f(x) e^{ikx} \, dx = (AL/\sqrt{2\pi}) \cdot \mathrm{sinc}(kL/2) \quad (5.10)$$

where k represented a spatial *angular* frequency, and had nothing to do with light waves. This can also be written in terms of the spatial frequency, $s = k/2\pi$,

$$F(s) = (AL/\sqrt{2\pi}) \cdot \mathrm{sinc}(L\pi s) \quad (5.11)$$

With $L = 1$ and $AL = \sqrt{2\pi}$, this is plotted in Figure 5.4.

108 THE DIFFRACTION OF LIGHT AND FOURIER TRANSFORMS

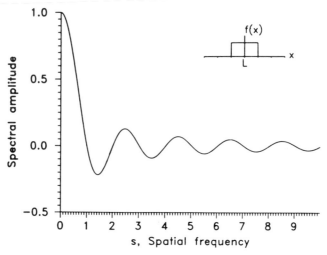

Figure 5.4 The positive frequency spectrum of our earlier (Section 4.2) one-dimensional spatial pulse of width L.

Now in $\text{sinc}(kL/2) = \text{sinc}(L\pi s)$ of (5.10) and (5.11), our one-dimensional problem, the zeros occur when

$$kL/2 = L\pi s = n\pi, \qquad n = 1, 2, 3, \ldots$$

at integral multiples of π. The zeros occur when

$$k = n2\pi/L; \qquad s = n/L$$

The first zero in the spectrum (5.11) occurs at a spatial frequency of

$$s = 1/L, \text{ or at the angular spatial frequency, } k = 2\pi/L.$$

Recall that the width of the square signal was L.

In our present two-dimensional diffraction expression for $E(X, Y)$, (5.8) or (5.9), we now have quantities kX/R and kY/R, where now $k = 2\pi/\lambda$, instead of just k in the exponents of the integrals. These kX/R and kY/R must therefore be equivalent to k, our one-dimensional spatial *angular* frequency. But frequencies where? *It is fundamental to our work to establish what these new spatial angular frequencies, kX/R and kY/R, represent,* so as to avoid confusion with our earlier mathematical use of k before light diffraction was even considered.

Before we continue, it will be helpful, now that we are working in two dimensions, to introduce some new notation. Let

$$u = \text{spatial angular frequency} = kX/R \quad (5.12a)$$
$$v = \text{spatial angular frequency} = kY/R \quad (5.12b)$$

From now on, working in two dimensions the use of k will be reserved for $k = 2\pi/\lambda$, with λ the wavelength of the light being diffracted. We have adopted u and v for our new spatial angular frequencies. Likewise we need to adopt new symbols for the companion new spatial frequencies, ξ and η. That is,

$$u = 2\pi\xi \quad (5.13a)$$
$$v = 2\pi\eta. \quad (5.13b)$$

These are exactly analogous to our earlier temporal $\omega = 2\pi f$ and our earlier spatial $k = 2\pi s$, our earlier one-dimensional forms.

> **But from this point on, k will always represent $2\pi/\lambda$ of the light being diffracted.**

Then let us assume that our aperture of (5.9) is a square of size L by L. We have already solved the integral. In its present form there will result, separately for the X and Y functions,

$$F(X) = (AL/\sqrt{2\pi}) \cdot \text{sinc}[(L/2)kX/R] = (AL/\sqrt{2\pi}) \cdot \text{sinc}(kL/2R)X$$
$$F(Y) = (AL/\sqrt{2\pi}) \cdot \text{sinc}[(L/2)kY/R] = (AL/\sqrt{2\pi}) \cdot \text{sinc}(kL/2R)Y$$

We write each in the second form since they are displayed in the Fourier X–Y plane.

The complete solution of (5.9) is the product of these two functions, (5.15). Now that we have recognized new spatial angular frequencies, $u = kX/R$ and $v = kY/R$, we can rewrite our solution as a spatial spectral function of these frequencies, u and v, (5.16). Or we can write it in terms of the new spatial frequencies, $\xi = u/2\pi$ and $\eta = v/2\pi$, by appropriately substituting ξ and η for u and v and simplifying, (5.17).

$$\xi = u/2\pi = kX/R2\pi \quad (5.14a)$$
$$\eta = v/2\pi = kY/R2\pi \quad (5.14b)$$

$$E(X, Y) = (A^2L^2/2\pi) \cdot \text{sinc}[(kL/2R)X] \cdot \text{sinc}[(kL/2R)Y] \quad (5.15)$$

$$E(u, v) = (A^2L^2/2\pi) \cdot \text{sinc}[(L/2)u] \cdot \text{sinc}[(L/2)v] \quad (5.16)$$

$$E(\xi, \eta) = (A^2L^2/2\pi) \cdot \text{sinc}(L\pi\xi) \cdot \text{sinc}(L\pi\eta) \quad (5.17)$$

Figure 5.5 is a graphical representation of equation (5.17), for $L = 1$ and $A^2L^2/2\pi = 1$. The zeros of these functions occur when

$$(kL/2R)X = (L/2)u = L\pi\xi = n\pi$$
$$(kL/2R)Y = (L/2)v = L\pi\eta = n\pi$$

The zeros occur at:

$$X = n2R\pi/kL \qquad u = n2\pi/L \qquad \xi = n/L$$
$$Y = n2R\pi/kL \qquad v = n2\pi/L \qquad \eta = n/L$$

Notice the linearity between n and X, n and u, n and ξ. Double the frequency means double the X. The first zero, $n = 1$, occurs at

$$X = 2R\pi/kL \qquad u = 2\pi/L \qquad \xi = 1/L$$
$$Y = 2R\pi/kL \qquad v = 2\pi/L \qquad \eta = 1/L$$

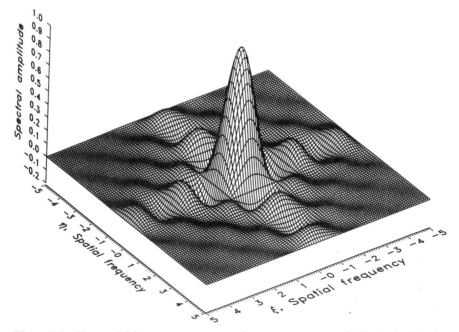

Figure 5.5 The spatial frequency spectrum of a square aperture of side L; the Fourier transform of the square function. Spectrum amplitude is normalized to unity at the origin.

> The aperture is a square of dimensions *L* by *L*.

> There is an exact analogy in the Fourier transformations; here dealing with two-dimensional Fraunhofer diffraction of light, and with the strictly mathematical, one-dimensional case, described earlier. *L* was the width of the one-dimensional mathematical pulse, and here *L* is the size of the square aperture.

5.4 DIFFRACTION APPLICATIONS OF THE NEW NOTATION

The manner in which the light has been spread out, or scattered, by the aperture, that amplitude pattern *represents* the Fourier transform of the aperture function. Diffracting visible light, it is not possible to see or to detect this amplitude pattern. What is actually visible is the square of this function; it is called the *irradiance*, and we display it (the relative irradiance) in Figure 5.6, the square of Figure 5.5. The square aperture function itself can also be plotted (Figure 5.7).

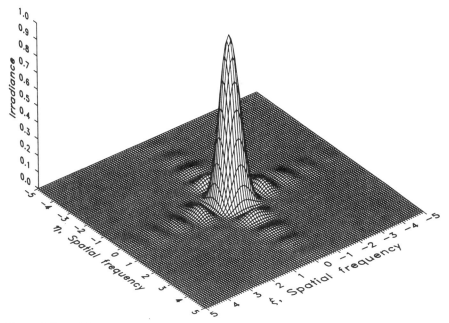

Figure 5.6 While Figure 5.5 represents the instantaneous light amplitude pattern in the Fraunhofer diffraction plane, what would be seen is this irradiance pattern. It is the square of Figure 5.5, again, normalized to unity at the origin.

112 THE DIFFRACTION OF LIGHT AND FOURIER TRANSFORMS

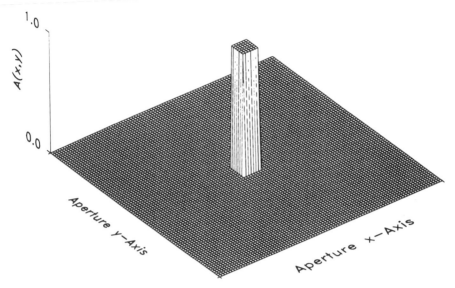

Figure 5.7 Figure 5.5 is the Fourier transform of this square function.

For a square (aperture) that has sides equal to one ($L = 1$) and a flat top (aperture illuminated by a plane wave) Figure 5.5 is its normalized Fourier transform, its two-dimensional spatial frequency spectrum. The spatial frequency spectral content here is specifically dictated by the aperture function shape, Figure 5.7, and by $L = 1$. Hence the spectral content is displayed in the Fraunhofer diffraction pattern, in the Fourier transform plane, but it corresponds to structures in the aperture plane, here a square aperture uniformly illuminated with monochromatic light, i.e., light of a single wavelength, ($\lambda = 2\pi/k$). The level of illumination on the aperture sets the height of the aperture function (Figure 5.7). A level of unity (one) is convenient.

A comment is appropriate here concerning the observation that $E(X, Y)$ takes on negative values in certain regions. Everywhere in these regions all the $\mathscr{E}\, dS$ contributions from the aperture have summed to be out of phase with respect to the light in the positive regions of $E(X, Y)$. Recall that cosine-type waves are propagating from the various dS in the aperture to points in the Fourier transform (Fraunhofer diffraction) plane. Over one complete cycle a cosine wave is $\frac{1}{2}$ positive valued and $\frac{1}{2}$ negative valued. Waves from various dS of the aperture get out of step (out of phase) with one another because of the different distances they must travel to a point $P(X, Y)$ in the transform plane. The sum of all the dE contributions at a point $P(X, Y)$ must therefore either be positive, negative, or zero. This is the identical phase effect which gives rise to interference fringes in Thomas Young's double-slit experiment. Kaiser and Russell (1980) have described an experiment to illustrate this phase effect and others related to it.

DIFFRACTION APPLICATIONS OF THE NEW NOTATION 113

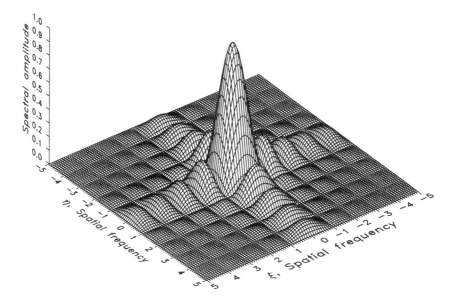

Figure 5.8 The absolute value of the amplitude function is often used to illustrate the spectrum, but clearly now, the phase information is missing.

Because the amplitude spectrum (Figure 5.5) goes negative in some regions, making it a little difficult to compare relative amplitudes, some people prefer to display the modulus, or absolute value, of the spectrum. We show that here in Figure 5.8.

5.4.1 Using the Angles of Diffraction

The light is diffracted at angles to the optical Z-axis and we can write the spectral functions in terms of these angles. Let the *angles of diffraction* be θ in the X-direction, ϕ in the Y-direction. Then, as shown in Figure 5.9,

$$X/R = \sin\theta, \qquad Y/R = \sin\phi \tag{5.18}$$

and (5.15) would become

$$E(\theta, \phi) = (A^2 L^2/2\pi) \cdot \operatorname{sinc}\left(\frac{kL \sin\theta}{2}\right) \cdot \operatorname{sinc}\left(\frac{kL \sin\phi}{2}\right) \tag{5.19}$$

114   THE DIFFRACTION OF LIGHT AND FOURIER TRANSFORMS

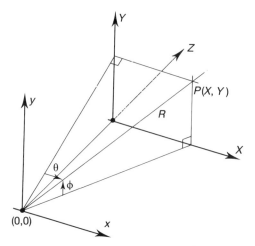

Figure 5.9 A little geometry allows the expression of these patterns in terms of diffraction angles rather than X, Y coordinates. For all practical purposes R can be considered constant.

where, of course, $k = 2\pi/\lambda$ (λ = wavelength of the light used). Zeros occur when

$$\left.\begin{array}{l} \dfrac{kL}{2} \sin\theta = \dfrac{2\pi L}{\lambda 2} \sin\theta = n\pi \\[2mm] \dfrac{kL}{2} \sin\phi = \dfrac{2\pi L}{\lambda 2} \sin\phi = n\pi \end{array}\right\} \quad (n = 1, 2, 3, \ldots)$$

Zeros occur when

$$n\lambda = L\sin\theta \quad \text{and} \quad n\lambda = L\sin\phi$$

Either of these is the familiar diffraction equation associated with the minima (the zeros) in the diffraction pattern of the one-dimensional single slit, L wide. Here the value of L is the extreme width of the aperture in the x- and y-directions. Since L and $\sin\theta$ are inversely related through $n\lambda = L\sin\theta$, large L in an aperture causes diffraction spread at small angles, θ; small L causes diffraction spread to larger angles, θ. The large L space corresponds to a small $1/L$ spatial frequency in the *aperture function* and first zero, $n = 1$, at small angle

$$\theta_{n=1} = \arcsin(1 \cdot \lambda/L) \quad \text{and} \quad L = \lambda/\sin\theta_{n=1}$$

$$\theta_n = \arcsin(n\lambda/L) = \arcsin(n2\pi/kL) \tag{5.20}$$

DIFFRACTION APPLICATIONS OF THE NEW NOTATION

Thus the second zero occurs when $n = 2$ and corresponds to a space $= L/2$ and a spatial frequency in the *aperture function* of $2/L$ and a larger angle of diffraction,

$$\theta_{n=2} = \arcsin(2 \cdot \lambda/L) \quad \text{and} \quad L = 2\lambda/\sin\theta_{n=2}$$

This idea is the basis of frontier research in diffraction analysis of structures, e.g., measure the angle of diffraction to the first zero and "Bingo!" you have the largest structure. Instead of using diffraction to measure λ of the light, we know λ of the light; we measure the diffraction pattern to determine the structure which caused it. This is sometimes referred to as an "inverse problem." You can understand why.

Our example was actually light diffracted at angles θ and ϕ, so it might be useful to plot the diffraction pattern as a function of functions of θ and ϕ. In Figure 5.5, ξ and η ranged from -5 to $+5$. With $L = 1$ we plotted $\text{sinc}(L\pi\xi) \cdot \text{sinc}(L\pi\eta)$ and it took us out exactly to the fifth zero, $n = 5$. From (5.14a)

$$\xi = kX/R2\pi = (k\sin\theta)/2\pi$$

$$\sin\theta = \xi(2\pi/k)$$

$$\text{sinc}[L\pi(k\sin\theta)/2\pi] \cdot \text{sinc}[L\pi(k\sin\phi)/2\pi]$$

The quantities $(kL\sin\theta)/2\pi$ and $(kL\sin\phi)/2\pi$ range between ± 5. We replot the function, again $L = 1$, and relabel the axes as $\sin\theta$ and $\sin\phi$ in units of $2\pi/kL$ (Figure 5.10). Hence the plot extremes of $\sin\theta$ and $\sin\phi$ are $\pm 5(2\pi/kL)$.

You can see that to actually determine the diffraction angles one must specify λ in the same units as L ($= 1$). If our square aperture was of $L = 1.0$ mm, then, using (5.20), the first zero occurs at

$$\theta = \sin^{-1}(1 \cdot (2\pi/k)) = \sin^{-1}(2\pi)/(2\pi/\lambda) = \sin^{-1}(\lambda) \quad (\lambda \text{ in mm})$$

Using laser light of wavelength $\lambda = 632.8 \times 10^{-9}$ m $= 632.8 \times 10^{-6}$ mm, the first zero is at an angle of

$$\theta = \sin^{-1}(0.0006328) = 0.03626 \text{ radians}$$

$$\theta = (0.3626 \text{ radians}) \times (180°/\pi \text{ radians}) = 2.077°.$$

The amplitude pattern of the diffracted light in the X–Y plane *represents* the Fourier spectrum of the aperture function. When we look at $E(\xi, \eta)$ of Figure 5.5, the spatial frequency spectrum is that which makes up the structure of Figure 5.7, $A(x, y)$. In Chapter 7 we will illustrate the back and forth transformation between such Fourier transform pairs. In the X–Y plane, with $k = 2\pi/\lambda$, the relationships are, again,

$$u = kX/R = (2\pi/\lambda)(X/R) = 2\pi\xi$$

$$v = kY/R = (2\pi/\lambda)(Y/R) = 2\pi\eta$$

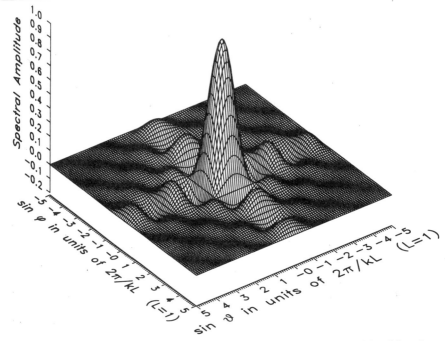

Figure 5.10 The Fourier transform can be plotted in terms of the sines of the diffraction angles. Here, $L = 1$.

5.4.2 The Rectangular Aperture and Four Forms of the Fourier Transform

Given a rectangular aperture $A(x, y)$ of uniform amplitude transmission, of width a along the x-axis, width b along the y-axis, its Fourier transform (its spectrum), except for other constant factors, is given by:

$$\text{Form of the Fourier Transform of } A(x, y) = \begin{cases} E(X, Y) = (Aab)\,\text{sinc}(akX/2R)\,\text{sinc}(bkY/2R) & (5.21a) \\ E(u, v) = (Aab)\,\text{sinc}(au/2)\,\text{sinc}(bv/2) & (5.21b) \\ E(\xi, \eta) = (Aab)\,\text{sinc}(a\pi\xi)\,\text{sinc}(b\pi\eta) & (5.21c) \\ E(\theta, \phi) = (Aab)\,\text{sinc}\left[\dfrac{a\pi \sin \theta}{\lambda}\right]\text{sinc}\left[\dfrac{b\pi \sin \phi}{\lambda}\right] & (5.21d) \end{cases}$$

As the y-width of the single slit is made twice the x-width, $b = 2a$, its diffraction pattern keeps its form but shrinks in toward the origin, exactly as shown along the η-axis, comparing Figures 5.8 and Figures 5.11 or 5.12. And indeed, the first zero in η which occurred at $\eta = 1$ before the doubling, now shows up at $\eta = \tfrac{1}{2}$ in Figures 5.11 and 5.12. Notice in Figures 5.12 and 5.8 that no change occurred in the ξ-direction.

The *functional form* of the Fourier transform of *all* rectangular apertures with uniform amplitude transmission will be the forms of (5.21). This is a form

DIFFRACTION APPLICATIONS OF THE NEW NOTATION   117

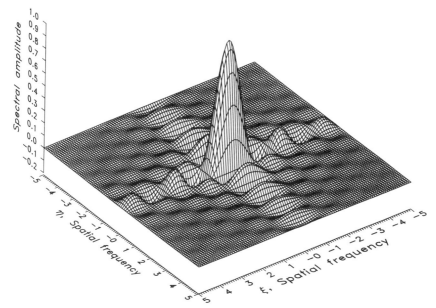

Figure 5.11   The representation of the Fourier transform of a rectangular aperture function. The x-width is half that of the y-width, hence greater spreading of the pattern in ξ.

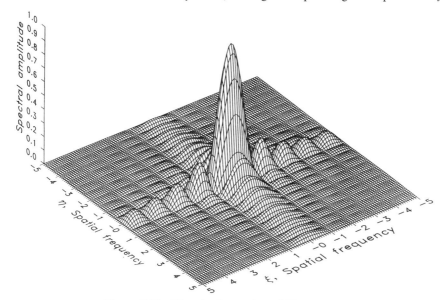

Figure 5.12   The absolute value of Figure 5.11.

which could be written as $(\text{sinc } \alpha)(\text{sinc } \beta)$, again defining sinc α as $(\sin \alpha)/\alpha$. But since we have so many different ways of writing the variables α and β it is worthwhile to compare them, say along the X-axis, the direction parallel to the aperture a-width on the x-axis (Figure 5.13).

118 THE DIFFRACTION OF LIGHT AND FOURIER TRANSFORMS

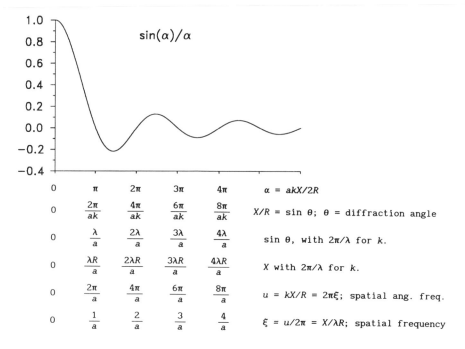

Figure 5.13 The Fourier transform of a pulse of width $= a$, displays nicely in terms of spatial frequency, ξ. But other quantities could be used if desirable.

Seeing these relationships we can understand that the Fourier transforms which provide E could be written several ways. All have "understood" constants before the integral.

1. In terms of actual coordinates in the X, Y plane:

$$E(X, Y) = \iint_{\text{aperture}} A(x, y) e^{ik(Xx + Yy)/R} \, dx \, dy \quad (5.22)$$

2. In terms of angular spatial frequencies $u = kX/R$ and $v = kY/R$:

$$E(u, v) = \iint_{\text{aperture}} A(x, y) e^{i(ux + vy)} \, dx \, dy \quad (5.23)$$

which is a nice simple expression. This is sometimes referred to as the "angular" spectrum because of the use of angular spatial frequencies.

3. In terms of spatial frequencies $\xi = u/2\pi$ and $\eta = v/2\pi$:

$$E(\xi, \eta) = \iint_{\text{aperture}} A(x, y) e^{i2\pi(\xi x + \eta y)} \, dx \, dy \quad (5.24)$$

4. In the small angle approximations,

$$X/R = \sin\theta \cong \theta \quad \text{and} \quad Y/R = \sin\phi \cong \phi,$$

where θ and ϕ are the angles of diffraction. We can write E in terms of these angles (or their sines).

$$E(\theta, \phi) = \iint_{\text{aperture}} A(x, y) e^{ik(\theta x + \phi y)} \, dx \, dy \quad (5.25)$$

This also is sometimes referred to as the "angular" spectrum because of the use of angles of diffraction (or their sines).

In a physics course which does not utilize the concept of the Fourier spectrum, but deals mainly with the geometry of diffraction, form (5.22) alone would probably be used. In an electrical engineering course concerned with propagation of radiation from antennas, form (5.25) might be most useful for understanding the radiated far field and such concepts as directivity, radiation patterns, beam width, and side lobes. See the references at the end of this chapter.

For someone interested in electron or x-ray diffraction, form (5.24) is useful because the spatial frequencies ξ and η refer directly to the reciprocals of the dimensions of the diffracting structures, e.g., to the spiral size and shape of DNA or to the "size" of the electron density distribution for a single molecule of, say, cyclopropane-carbo-hydrazine. For the purposes of the present work we find form (5.24) most useful because of this direct reference to the dimensions of the optical aperture causing the diffraction. For its simplicity we will sometimes use form (5.23), using the angular spatial frequencies, u and v, perhaps changing to spatial frequencies ξ and η after the mathematics is done. We will return to form (5.22) in Section 5.6 where we solve the problem related to the required large value of R.

The vertical axis then, representing $E(X, Y)$, a point function and usually normalized to unity at the origin, is the relative scale of the *amplitude* of each Fourier spatial spectral component in units of $[\text{Watts/m}^2]^{1/2}$. This quantity cannot be physically seen or, at optical wavelengths, detected. It corresponds to the scalar amplitude of the light passing through the Fourier transform plane (the $X-Y$ plane). What is easily detectable is a measure of the square of this quantity and, when measured over a finite area of the $X-Y$ plane, is the time-averaged rate of flow of radiant energy through that plane, the radiant flux density or irradiance in units of Watts/meter^2, commonly called, in the past, "intensity."

$[E(X, Y)]^2$ measured in the diffraction plane *could* be referred to as "surface power density" with units of [Watts/meter^2] or [Watts/area increment, $\Delta X \Delta Y$]. If we look at it as $[E(\xi, \eta)]^2$ it could be called "spatial frequency spectral power density" or "power spectrum of spatial frequencies" with units of [Watts/spatial frequency increment, $\Delta\xi\Delta\eta$] or [Watts·meter^2], using an appropriate conversion factor to change meter^{-2} to meter^{+2}.

5.5 SCALE IN THE FOURIER TRANSFORM PLANE

It has been suggested, after Equation 5.20 that in the diffraction plane (Fourier transform plane) the position of the first zero along, say, the X-axis of the diffraction pattern, is related to the largest x-axis opening in the aperture; and in fact if that opening is "a" wide, the spatial frequency coordinate ξ of the first zero will be $1/a$ [cycles/meter]. If $a = 1$, the first zero is at $\xi = 1$ [cycle/meter]. In our simple examples the light diffracted (directed by diffraction) to the position of the first zero is diffracted by a completely open structure of width a. All dE there sum to zero in our simple examples. This may be true for apertures of uniform transmittance, symmetry, and of simple geometrical shape, but due to phase and nonsymmetry, in general, one cannot rely on an actual zero being present at that position. Apertures can have a variety of structures and transmittances (and phases) and although the largest, completely open structure would cast a zero where we expect it, other structures might diffract light also to the same position, perhaps light diffracted by aperture structures of phase, random orientation, and non-uniform transmittance. Except for simple apertures, we should exercise care in our assumptions about deducing $A(x, y)$ structure from $[E(X, Y)]^2$ measurements.

Going in the other direction, how does one construct the *spatial frequency scale*, for ξ and η, for a diffraction pattern in the $X-Y$ plane, knowing the aperture maximum extent? For simple apertures, even though the diffraction pattern may be complicated, we do have certain information; we know R, λ, and we know the maximum extent of the aperture in, say, the x-direction; let that be $a = 0.01$ m, let $\lambda = 600$ nm and let $R = 2000$ m. In the $X-Y$ plane, at what distance from the origin along the X-axis will light be diffracted due to this 0.01 m aperture structure? That position corresponds to the first zero caused by spatial frequencies of $1/0.01$ m, i.e., 100 [spatial cycles/m] in $A(x, y)$. We see from (5.21a) that the X distance corresponding to the first zero is given by $akX/2R = 1 \cdot \pi$, or $X = \lambda R/a$. Here,

$$X = \frac{(600 \times 10^{-9} \text{ m})(2000 \text{ m})}{10^{-2} \text{ m}} = 0.12 \text{ m}$$

Along the X-axis, 0.12 m from the center of the diffraction pattern is the position of the first zero caused by 0.01 m spatial structures in the aperture. In spatial frequency terms we label this point $1/0.01 = 100$ [cycles/meter]. It is along the X-axis, 0.12 m from the origin. Then, because of the linearity between X and ξ (5.14a), 0.24 m along the X-axis corresponds to 200 [cycles/meter], etc., and the axes of the diffraction pattern can be labeled as a spatial frequency spectrum. Of course the relationships between ξ and X, and between η and Y, are explicit in equations (5.14a) and (5.14b).

All our subsequent calculations use, in effect, a large value of R. What we have shown in this chapter is that the $E(X, Y)$ function, the amplitude of the light distribution in a plane perpendicular to the optical axis at large R from the aperture, is given by the two-dimensional Fourier transform of the aperture function. Hence this plane is referred to as the Fourier transform plane. The $E(X, Y)$ function can be called the spatial frequency spectrum of the aperture function, if we think of it in terms of ξ and η. The irradiance function, $[E(X, Y)]^2$, is what one sees or measures under Fraunhofer diffraction conditions. The observed Fraunhofer diffraction pattern is a representation of (is directly proportional to) the square of the Fourier transform of the aperture function $A(x, y)$.

5.6 MAKING THE FOURIER TRANSFORM ACCESSIBLE AT REASONABLE DISTANCE

With a plane parallel beam of monochromatic light incident upon a small aperture, the diffraction pattern some 1000 meters or more further along the optical axis, *will not be*, but "*will contain*," a very good approximation of the Fourier transform of the aperture function, $A(x, y)$. It is only a good approximation because there will be at the observation plane (the Fourier transform plane) some slight phase variation which would not be present in the true Fourier transform of $A(x, y)$. Banerjee and Poon (1991) refer to it as "phase contamination" (on our Fourier transform), a nice way of putting it. A portion of the phase contamination is due to the approximations we made. The contamination is small but it can be important in some situations. Mathematically, such phase variation in the transform plane, $\Phi(X, Y)$, would be incorporated into the expression for the optical disturbance in the transform plane as an exponential factor,

$$e^{i\Phi(X,Y)}$$

This is a complex number. Note: we use capital phi in the transform plane. The units of $\Phi(X, Y)$ would typically be the dimensionless radians, describing phase variation in the transform plane. At optical wavelengths the only optical signal which is detectable is the *irradiance*, formerly referred to as *intensity*, both with dimensions of Power/area; and irradiance is proportional to the square of the optical disturbance. Hence, whatever we detect, see, photograph, measure, etc., in the Fourier transform plane, the irradiance there will be proportional to the square of the optical disturbance. Squaring the phase function factor means multiplying it by its complex conjugate:

$$e^{i\Phi(X,Y)} \cdot e^{-i\Phi(X,Y)} = e^{i\Phi(X,Y) - i\Phi(X,Y)} = e^0 = 1.$$

The irradiance pattern in the Fourier transform plane will be proportional to the **square** *of the Fourier transform of the aperture function $A(x, y)$. All evidence of* **any** *extraneous variation of phase from the true transform will not be apparent in the irradiance.* **If the only quantity we wish to observe is irradiance in the transform plane then phase contamination may not be much of a problem!**

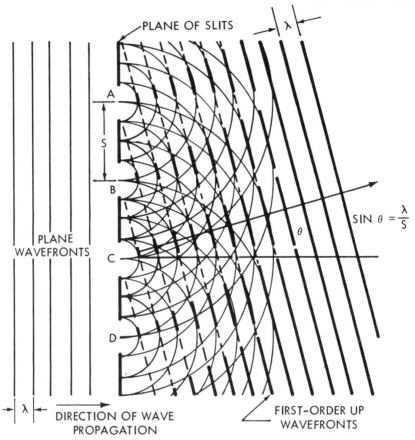

Figure 5.14 An aperture of slits diffracting light at angle $\theta = \sin^{-1}(\lambda/S)$. (From A. R. Shulman, *Principles of Optical Data Processing for Engineers*, NASA Technical Report R-327, May, 1970.)

Figure 5.14 shows a diffracting aperture with plane parallel monochromatic light incident upon it. A particular spatial frequency structure is diffracting light at its particular angle of diffraction, θ (this angle is exaggerated). Some 1000 meters or more further down the optical axis will be a plane in which we would be able to measure and see an irradiance pattern which is a good representation of the square of the Fourier transform of the aperture function. At this large distance, R, there still exists a small but negligible phase contamination, but it is not apparent because we see or detect only the complex square of the optical disturbance. The approximations making this true are the Fraunhofer approximations discussed in Section 5.2. We used a large R so as to make phase contamination negligible.

In the focal plane of a thin lens placed after the diffracting aperture we see that of *all* the rays diffracted at a particular angle (the angle is exaggerated in

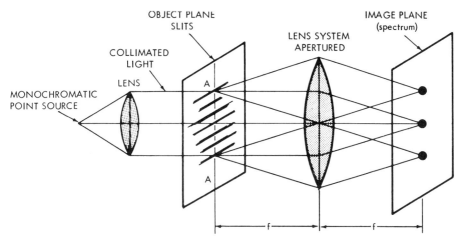

Figure 5.15 Examining either ray which leaves the slits at angle, say θ, and passes "undeviated" through the lens center, the angle from the lens to the spectrum plane remains θ. The Fourier transform can be displayed in the focal plane of a lens following the diffracting aperture. (From A. R. Shulman, *Principles of Optical Data Processing for Engineers*, NASA Technical Report R-327, May, 1970.)

Figure 5.15) due to a particular spatial frequency structure, *all* are parallel and are brought to focus at a single point in the focal plane of the lens. By examining the angle of the *one particular* diffracted ray which passes "undeviated" through the center of the thin lens you can observe that this angle is unchanged from that without the lens. The Fourier transform pattern has now been transferred from large distance R to the lens focal plane at distance F. Won't that be convenient! (A diagram such as Figure 5.15 is often drawn incorrectly by book illustrators.)

But we never get something good without paying a price! Recall there was a slight and probably negligible phase contamination in the Fourier transform at distance R. Now, in the lens focal plane where the optical disturbance will be formed by spherical waves converging from the lens (that's what a lens does), there will be an additional significant phase variation in the Fourier transform in the form of phase curvature. The optical disturbance will represent the Fourier transform, but it will not be on a plane but on an almost spherical surface. So, except for our earlier and negligible phase contamination, a very good Fourier transform of the aperture will be present on a spherical surface. The vertex of this spherical surface is on the optical axis at the rear focal point of the lens. If the lens has long focal length, the phase curvature is not large, and if we consider only regions in the proximity of the optical axis, detecting only intensity, the curvature matters little. We will elaborate on the phase curvature and how to get rid of it in Chapter 6, Section 6.1.3.

Equation (5.8) illustrated our expression for the Fourier transforms which would be found in a plane a distance R, some 1000 meters or more, after the diffracting aperture. The quantities X/R and Y/R provided the tangents of the diffraction angles. *With a lens those angles can be unchanged* but now the tangents can be expressed as X/F and Y/F, with X and Y, again capitals, being coordinates in the focal plane, and F the focal length of the lens. Some authors phrase it such that the lens creates a Fourier transform in its focal plane. We rewrite Equation (5.8) as (5.26), using F in place of R.

$$E(X, Y) = \iint_{\text{aperture}} A(x, y) \cdot e^{ik(Xx + Yy)/F} \, dx \, dy \quad (5.26)$$

Thus, in the focal plane, $X-Y$, of a lens suitably placed after a properly illuminated aperture, say a square aperture, there will appear an irradiance pattern which represents the complex square of the Fourier transform of the aperture function (see Figure 5.16). As we have done in this chapter we will reorient such figures in subsequent chapters.

Using the conversion equations of this chapter we know we can rescale the X- and Y-axes to represent the spatial frequencies ξ and η, replacing R with F.

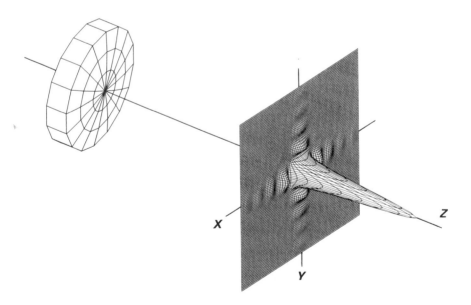

Figure 5.16 Under Fraunhofer conditions, if the $X-Y$ plane is the focal plane of a lens following a diffracting aperture, that's where we'll find the Fraunhofer irradiance pattern, the square of the Fourier transform of the aperture function. Can you tell what was the aperture shape?

We rewrite (5.14a) and (5.14b):

$$\xi = u/2\pi = kX/F2\pi \quad (5.27a)$$

$$\eta = v/2\pi = kY/F2\pi \quad (5.27b)$$

After the conversion process there will be present in the rear focal plane of the lens the following function, subject to all the stated qualifications.

$$E(\xi, \eta) = \iint_{\text{aperture}} A(x, y) \cdot e^{i2\pi(\xi x + \eta y)} \, dx \, dy \quad (5.28)$$

Is it true, that a university is like a Fraunhofer aperture, in that, students passing through, experience a four-year transformation?

5.7 THE "AMPLITUDE" OF OPTICAL DISTURBANCES

If constant over a reasonable size area, the magnitude of $E(X, Y)$, the amplitude of an optical disturbance, can be related to measurable quantities,

$$E^2 = I = \langle S \rangle = cU/2 = [(c\varepsilon_0/2)E_0^2] \quad (5.29)$$

where:

- I = irradiance on the area, measurable in Watts/meter^2,
- $\langle S \rangle$ = time average of the Poynting vector magnitude, S, of the incident light in Watts/meter^2,
- c = speed of light in meter/second,
- U = energy density, radiant energy per unit volume in Joules/meter^3,
- ε_0 = electric permittivity of free space in Farad/meter,
- E_0 = amplitude of the electric field of the incident light, in Newtons/Coulomb.

Curiosity leads us to ask about the MKS units of $\mathscr{E}(x, y)$, the amplitude per unit area of the optical disturbance emergent from the aperture, referred to in Section 5.1. Few authors, if any, comment on this. Obtaining the units will then allow us to see the dimensions of $\mathscr{E}(x, y)$.

To represent the instantaneous amplitude of a spherical wave from an idealized point source we can use an expression such as

$$\psi(r, t) = (\mathscr{A}/r) e^{ik(r \mp vt)}, \quad r \neq 0$$

In optics the point source is a source of radiant power, Watts. We know that at some distance from the radiant point source the irradiance describes the power flux in Watts/m^2. The irradiance is proportional to ψ^2 or $\psi \cdot \psi^*$ and hence proportional to $(\mathscr{A}/r)^2$. Hecht (1987) calls $\mathscr{A}$ the "source strength." $(\mathscr{A}/r)^2$ must have units of Watts/m^2, thus the MKS units of $\mathscr{A}$ must be Watts$^{1/2}$. This is a strange concept which nevertheless finds use in optics with *idealized* spherical waves from an *ideal* point source. The MKS units of ψ, the instantaneous scalar wave amplitude, are [Watts/m^2]$^{1/2}$.

For diffraction phenomena in general an extended concept arises. One often begins with Huygen's principle utilizing point sources in a diffracting aperture. An expression is constructed to find the optical disturbance in a diffraction plane at some distance further down the optical axis. Many point sources are grouped together in a differential area dS of the aperture; the linear extent of $dS \ll \lambda$, all the point emitters within dS are in phase, and dS is essentially a point. The optical disturbance in the diffraction plane attributed to the many point emitters within dS we call dE. dE is often set up as

$$dE = (\mathscr{E}_A/r) \cdot e^{i(\omega t \pm kR)} \, dS \qquad (5.30)$$

Hecht calls $\mathscr{E}_A$ the source strength per unit area; it is identical to the $\mathscr{E}(x, y)$ we used in (5.1) with units of [amplitude/m^2]. dE represents a scalar amplitude in the diffraction plane. dS is so small, the dE refer to spherical wave fronts, hence the amplitude decreases as $1/r$.

After integration of all dE over all dS, $E(X, Y)$ (a point function) in the diffraction plane is obtained. E^2 yields the measurable irradiance function. E^2 and $(dE)^2$ must have MKS units of Watts/m^2. From the differential expression (5.30), the units of dE are [Watts/m^2]$^{1/2}$. The units of $\mathscr{E}_A$, the source strength per unit area, are Watts$^{1/2}$/m^2. If $\mathscr{A}$ is the source strength of a single point source with units Watts$^{1/2}$, then indeed $\mathscr{E}_A$, the source strength per unit area should have units of Watts$^{1/2}$/m^2. In this model, $\mathscr{E}_A$ would be equal to $\mathscr{A}$ multiplied by the number of point sources per unit area.

It is difficult to assign a physical meaning to Watts$^{1/2}$. This curiosity has arisen because, of all classes of optical theory, scalar wave theory is one or two classes away from the truth. One does not encounter Watts$^{1/2}$ using electromagnetic or quantum theory of optics. We may calculate the time averages of Poynting's vector or photon flux but we don't believe there has ever been assigned a physical meaning to their square roots. In (5.29) we did not describe E but E^2. Scalar wave theory can describe many optical phenomena but "light" is in all likelihood, a particle.

There is an exact and perhaps more clearly understood analogy. Energy density in a non-divergent beam of light has units of Joules/m^3. Fluid density in a non-divergent flow has units of kilograms/m^3. Multiply both by their velocities and we obtain the power flux for the light, (Joules per second)/m^2, and the mass flux for the fluid, (kilograms per second)/m^2. Watts$^{1/2}$, or (Joules per second)$^{1/2}$, has the same lack of physical meaning as (kilograms per second)$^{1/2}$.

5.8 A PRELIMINARY LIST OF SOME PROPERTIES OF FOURIER TRANSFORMS

5.8.1 Symmetry Properties of Fourier Transforms (One Dimensional)

Function $f(x)$	Fourier transform, $F(u)$ or $F(\xi)$
Complex, no symmetry	Complex, no symmetry
Hermitian*	Real, no symmetry
Antihermitian*	Imaginary, no symmetry
Complex, even	Complex, even
Complex, odd	Complex, odd
Real, no symmetry	Hermitian
Real, even	Real, even
Real, odd	Imaginary, odd
Imaginary, no symmetry	Antihermitian
Imaginary, even	Imaginary, even
Imaginary, odd	Real, odd

* Hermitian = real part even, imaginary part odd; antihermitian = real part odd, imaginary part even (after Gaskill (1978)).

5.8.2 Linearity Property

If

$$A(x, y) \xrightarrow{\text{FOURIER TRANSFORMS TO}} E(\xi, \eta)$$

and

$$B(x, y) \xrightarrow{\text{FOURIER TRANSFORMS TO}} F(\xi, \eta)$$

and a and b are arbitrary constants, then

$$[a \cdot A(x, y) + b \cdot B(x, y)] \xrightarrow{\text{FOURIER TRANSFORMS TO}} [a \cdot E(\xi, \eta) + b \cdot F(\xi, \eta)]$$

The transform of the sum of two functions is the sum of their transforms.

5.8.3 Shifting Property

If

$$A(x, y) \xrightarrow{\text{FOURIER TRANSFORMS TO}} E(\xi, \eta)$$

and p and q are constants, possibly zero, then

$$A(x \pm p, y \pm q) \xrightarrow{\text{FOURIER TRANSFORMS TO}} e^{\pm i2\pi(p\xi + q\eta)} \cdot E(\xi, \eta)$$

128 THE DIFFRACTION OF LIGHT AND FOURIER TRANSFORMS

The Fourier transform of a shifted function is the Fourier transform of the unshifted function multiplied by an exponential phase factor having linear phase.

5.8.4 Scaling Property

If

$$A(x, y) \xrightarrow{\text{FOURIER TRANSFORMS TO}} E(\xi, \eta)$$

and a and b are real non-zero constants, then

$$A(ax, by) \xrightarrow{\text{FOURIER TRANSFORMS TO}} (1/ab) E(\xi/a, \eta/b)$$

and

$$a \cdot A(x, y) \xrightarrow{\text{FOURIER TRANSFORMS TO}} a \cdot E(\xi, \eta)$$

5.8.5 Fourier Transforms and Complex Functions

In general a Fourier transform is a complex-valued point function,

$$E(\xi, \eta) = \text{Re}\{E(\xi, \eta)\} + i \cdot \text{Im}\{E(\xi, \eta)\}$$

where the Re and Im functions are the mathematically real and imaginary components of $E(\xi, \eta)$. At each point (ξ, η), $E(\xi, \eta)$ can be complex; it can have a real part and an imaginary part. In exponential form, following the rules of complex algebra, the above expression becomes

$$E(\xi, \eta) = |E(\xi, \eta)| \cdot e^{i\Phi(\xi, \eta)}$$

The expression,

$$|E(\xi, \eta)| = \sqrt{[\text{Re}\{E(\xi, \eta)\}]^2 + [\text{Im}\{E(\xi, \eta)\}]^2}$$

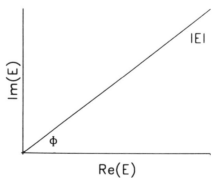

Figure 5.17 The relationships between $\text{Re}\{E(\xi, \eta)\}$, $\text{Im}\{E(\xi, \eta)\}$, $|E(\xi, \eta)|$, and $\Phi(\xi, \eta)$.

is called the spectrum (use the positive square root!) See Figure 5.17. In the mathematics of complex numbers, $|E(\xi, \eta)|$ is also called the modulus of $E(\xi, \eta)$. Everything is quite simple if the imaginary part of $E(\xi, \eta)$ is zero.

The following quantity,

$$\Phi(\xi, \eta) = \tan^{-1} \frac{\text{Im } E(\xi, \eta)}{\text{Re } E(\xi, \eta)}$$

is the phase function (see Figure 5.17).

The irradiance function $= E^2(\xi, \eta)$, also called the "power spectrum," is given by

$$|E(\xi, \eta)|^2 = [\text{Re}\{E(\xi, \eta)\}]^2 + [\text{Im}\{E(\xi, \eta)\}]^2 = E(\xi, \eta) \cdot E^*(\xi, \eta)$$

Taking square roots we see that the modulus of $E(\xi, \eta)$ can also be written as $\sqrt{E(\xi, \eta) \cdot E^*(\xi, \eta)}$, (positive root!).

GENERAL REFERENCES TO THE DIFFRACTION OF LIGHT

Champeny (1973), (1985); Ditchburn (1976); Duffieux (1983); Françon (1966); Guenther (1990); Heavens and Ditchburn (1991); Hecht (1987); Iizuka (1987); Klein and Furtak (1986); Nussbaum and Phillips (1976); Pedrotti and Pedrotti (1987); Steward (1983); Taylor (1978), (1987); Williams and Becklund (1972).

REFERENCES TO DIFFRACTION IN RADIO WAVE PROPAGATION

Balanis "field regions" (1982); Kraus (1984); Stutzman and Thiele (1981).

6

A BRIEF SUMMARY OF LINEAR SYSTEMS THEORY APPLIED TO OPTICAL IMAGING

The fact that a lens can produce the Fourier transform of an aperture or the transform of a two-dimensional scene within the aperture has great significance for optical imaging and, in general, for optical processing of two-dimensional structures. (And why not three-dimensional structures?) The necessary relationships needed to understand this significance have developed from the application of linear systems theory. In what follows we will not employ the full mathematical rigor of that theory; that is beyond the scope of this book. The details can be found in the references at the end of the chapter. Goodman (1968) did an excellent job applying linear systems theory to optical systems, and more recently, Banerjee and Poon (1991) seem to have made Goodman's work even more accessible. You are cautioned that some of the mathematics in the references is formidable. Let's see now if the essentials can be made understandable, without great mathematical detail, but with the mathematical concepts.

6.1 LINEAR SYSTEMS THEORY APPLIED TO OPTICAL SYSTEMS

6.1.1 A General Outline

Within linear systems, such as the optical systems we will consider, certain fundamental relationships exist. For brevity we present the following diagram, Figure 6.1, indicating those relationships. The arrows indicate the operation required to obtain one function from another. The symbol (*) means

LINEAR SYSTEMS THEORY APPLIED TO OPTICAL SYSTEMS   131

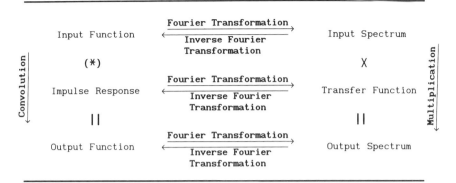

Figure 6.1 When optical systems can be treated as "linear systems," these relationships exist.

mathematical convolution. In the following pages we will describe the convolution operation and in Section 6.4 will show how this can be done with two-dimensional functions.

From what we already know about diffraction and Fourier transforms, the diagram strongly suggests that optical imaging systems may be describable with linear systems theory. *Under the proper conditions,* a single lens forming an image of an object is operating within the framework of a linear system, (Figure 6.2). For such an imaging device and with reference to Figure 6.1, the input

Figure 6.2 A linear system, in operation. Obviously a double exposure.

function will generally be a two-dimensional scene, e.g., a photo transparency; the output function will be the two-dimensional image. The lens, in performing this imaging function, transfers the spatial frequency information of the object plane to the image plane, with some alteration. The impulse response, the description of how the system images a single point (impulse) of monochromatic light, translates each point of the object plane to a geometrically appropriate point in the image plane. As the reader might surmise, there is a relationship between the minimum closeness of resolvable image points and the maximum spatial frequencies which can be imaged. We shall clarify the terminology shortly. Right now we refer only to the input and output functions and something called an impulse response, yet to be really defined.

6.1.2 Imaging

6.1.2.1 Coherent Imaging Let us immediately examine how an imaging system actually fits into this diagram. With the advent of optical lasers and their coherent light, two different kinds of imaging are now regularly done: imaging with coherent light, and imaging with incoherent light. Prior to the existence of lasers it was possible to obtain coherent light by concentrating a (usually bright) beam of light onto a pinhole aperture, and with a monochromatic filter, the light emerging from the pinhole was quite coherent. If a scene is illuminated with coherent laser light and we image it with a lens, |magnification| = 1, the following diagram is valid (Figure 6.3).

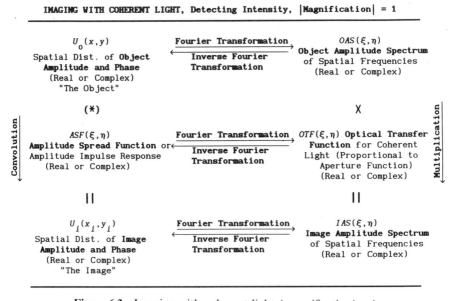

Figure 6.3 Imaging with coherent light, |magnification| = 1.

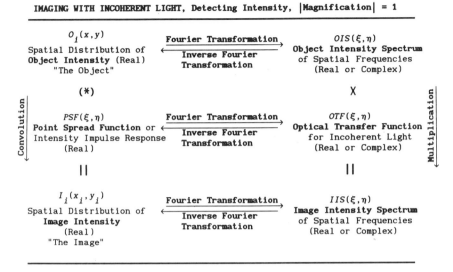

Figure 6.4 Imaging with incoherent light, |magnification| = 1.

Whether we have access to the amplitudes, U_o and U_i, in this scheme is beside the point. This is the way linear systems theory would describe this process. We still have not defined the amplitude impulse response, though its International Standard (ISO) acronym is likely to be *ASF* for amplitude spread function. We will use $ASF(\xi, \eta)$. We have yet to discuss the right side of Figure 6.3 but we do know something about Fourier transforms and inverse transforms—the transform pair; and we know something about spatial frequency spectra. But in Figures 6.3 and 6.4, the spread functions and transfer functions are mathematical constructs which describe what the optical system does to the light, transferring it from object to image.

6.1.2.2 Incoherent Imaging If we do more conventional imaging, e.g., with a telescope, a 35 mm camera or, say, with a video camera, with conventional incoherent (but monochromatic) illumination, Figure 6.4 is appropriate. Because the light is incoherent we cannot keep track of phase for any of it. So the best we can do is work with irradiances, formerly referred to as intensity.

Some readers may object to the use of the term "intensity" in this and subsequent diagrams. In the optical sense it has often been defined as dimensionally identical to the "irradiance," but irradiance is usually the more correct term to use. Linear systems theory owes much, if not most, of its development to the field of electrical engineering. In the 1991 Wiley book *Fundamentals of Photonics* (part of Wiley's Series in Pure and Applied Optics, written by B. E. A. Saleh and M. C. Teich, two senior professors of electrical engineering, Series Advisory Editor—Joseph Goodman, Past President of the Optical Society of America), the index does not mention the term "irradiance," though there are six entries referring to "intensity." In the list of symbols, "I" refers to optical intensity, with units of Watts/m^2. The universal replacement of the term "intensity" with the term "irradiance" is not yet complete. We are led to believe that "intensity" is to be reserved for quantities with SI units of Watts/steradian.

Although "irradiance" is perfectly correct for image descriptions, e.g., light falling on image screens, we are a bit uncomfortable when using it to describe what emerges from an object, especially one radiating by reflected light, where "radiant exitance," the old "radiant emittance" might be more proper. The interested reader might consult specific references cited at the end of this chapter.

In any case, in these diagrams the intensity will always mean the point function which describes dimensionally the average optical power per unit area at that point, emitted, reflected, irradiated, or just passing through. The distinction we are trying to make is between *amplitude* and *intensity*.

Notice the symmetry of the Fourier transformations and inverse transformations, and that across the middle of both diagrams is an impulse response and an optical transfer function, both also related by Fourier transforms. In both cases the imaging systems of concern here are those which process light waves, bringing them to divergence or convergence, and causing diffraction. From a scalar or vector wave theory of light we know that the net effect of two or more waves can be determined by a summation, i.e., utilization of the principle of linear superposition. Two distinctions should be made:

(1) **COHERENT LIGHT:** If the light waves to be added are derived from the same source such that there exists a fixed relationship between the phases of many of the waves, we say that coherence is present and the addition, the linear superposition, of *wave amplitudes* will indeed be meaningful. The concept of optical interference will come into play, even though here, as in Case (2), it will only be intensity, not amplitude, which is detectable in images at optical wavelengths.

(2) **INCOHERENT LIGHT:** If the waves are derived from effectively independent sources even though monochromatic, the phase relationships of the waves converging to the image plane will not be fixed, but will vary randomly; the net effect can only be determined by statistical means which is exactly what a light detector does. We find here that the average light irradiance (or intensity), *not amplitude*, is the only quantity describing net effects. This will require the linear superposition of irradiances (or intensities). We will come to understand that, because the image is formed by a convolution of an always positive object function and an always positive *SPREAD* function, it seems that completely destructive "*interference*" in the image function will be highly unlikely.

LINEAR SYSTEMS THEORY APPLIED TO OPTICAL SYSTEMS 135

We conjecture that nowhere in an image formed with incoherent light will the intensity be zero due to destructive interference.

With all forms of illumination at optical wavelengths the observable quantity is light power/unit area; up to the year 1995 we believe there have been no direct measuring detectors of the amplitude of visible light.

With incoherent illumination, Case 2, the system would be linear in light intensity, and hence in Figure 6.4 we work with object and image intensity functions.

With coherent illumination, e.g., laser light, Case 1, the system will be linear in light amplitude, and hence in Figure 6.3 we work with object and image amplitude and phase functions.

The input (object function) and output (image function) will be two dimensional as will be the spatial frequency spectra, the impulse responses, and the transfer functions. Additionally, many of these functions may be mathematically real or complex.

From the material of the previous chapter we know that a two-dimensional aperture can have a two-dimensional spatial frequency spectrum. In Chapter 5 we carried out such spectrum analysis by calculating the Fourier transform of two-dimensional apertures. In imaging we are concerned with two-dimensional objects and images, so it should not be surprising that we see in the two imaging diagrams, Figures 6.3 and 6.4, the Fourier transform operation capable of producing the spatial frequency spectra of objects and images. But in both imaging diagrams, across the middle row, we see two other functions come into play; they are a necessary part of linear systems theory: an *impulse response* and a *transfer function*. We need to know what these are and how they are determined.

6.1.3 Defining the Spread Functions

To help with the definitions we bring into play a third diagram (Figure 6.5); linear systems theory applied to Fraunhofer diffraction. Having examined Fraunhofer diffraction in the previous chapter we can see that Figure 6.5 is a complete description of that phenomena and apparently more. We know that at a large distance R from a properly illuminated aperture (real or complex), or near the focal plane of a lens of focal length F placed after the aperture, we can *find* (not see) the optical disturbance which represents the Fourier transform of the aperture function; that's the Fraunhofer amplitude; it cannot be seen or detected. What we would see or detect is its (complex) square, the Fraunhofer irradiance. If we do this at a large distance, R, we will *see* the irradiance function which represents the true square of the Fourier transform,

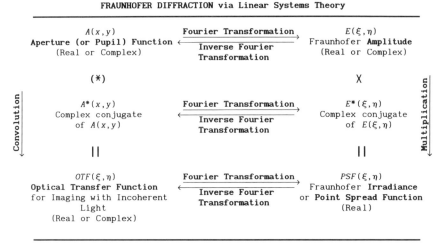

Figure 6.5 Fraunhofer diffraction via linear systems theory.

the larger R, the more true the square; the negligible phase contamination is not seen. This Fraunhofer irradiance is a correct representation of one form of a point spread function. But the transform present in the focal plane of a lens will not be true; there will be phase curvature, specifically the surface of constant phase is curved, nearly spherical.

When we calculate the Fourier transform of an aperture function, the calculation describes the phase of the transform in the $X-Y$ plane. When we optically produce a Fourier transform, there is only one position of the aperture for which the phase will be the true phase on the *plane* surface, a distance F from the lens; when the aperture is in the lens' front focal plane. At all other aperture positions the calculated transform phase is present on a curved surface whose vertex is at the lens' rear focal point. Depending on the aperture position, before or after the front focal plane, the phase curvature at the rear focal plane can be concave or convex. Goodman (1968) and others provide details.

If phase curvature exists we usually try to make it as small as possible. A long focal length lens is often chosen for this. In any case, mathematically, the phase curvature is represented in the transform as an exponential factor, $e^{i\Phi(X,Y)}$. The irradiance that we see contains the complex square of this exponential; it has vanished in the visible irradiance function.

Goodman (1968) describes the usefulness of placing the aperture *after* the lens; its distance from the lens changes the size of the transform. That should be useful, also!

The mere fact that a lens can form images, this writer sometimes considers wondrous. That a simple lens can also perform a two-dimensional Fourier transformation, in real time, immediately, (or at least at the speed of light), he considers no less amazing.

The square of the Fraunhofer amplitude (the amplitude impulse response) (multiply it by its complex conjugate if it is complex) is the "intensity impulse response," what we have termed the Fraunhofer irradiance or point spread function, *PSF*, (Figure 6.5). Viewing diffraction as a simple scattering of light in space, the dimensions of irradiance would be power/area-increment, Watts/$\Delta X \Delta Y$, in MKS units of [Watts/m^2]. But since we know that this pattern represents the square of the spatial frequency spectrum of the aperture function, its dimensions, rescaled, can be expressed as power/spatial-frequency-increment, Watts/$\Delta \xi \Delta \eta$, in 3-space in MKS units of [Watts·m^2], what we may term a spectrum of spatial frequency power. If viewed only in 2-space the power/spatial-frequency-increment, e.g., Watts/$\Delta \xi$, would have MKS units of [Watts·m].

In Figure 6.5 for Fraunhofer diffraction, two functions are named the same as those in the *imaging* diagrams, (Figures 6.3 and 6.4), the *point spread function* and the *optical transfer function*. You might well wonder why they are called that since in Fraunhofer diffraction no "imaging" occurs. When using a lens to form images, think where objects and images are located; it seems that all the requirements for Fraunhofer diffraction are being violated. Objects are often in close proximity to imaging lenses, as are the images. Imaging seems to be a phenomena carried out in the region where Fresnel diffraction occurs. Indeed this is the case. We will try to clarify this.

Consider a lens to be used for a specific imaging operation. There will be an object plane and an image plane. We can look upon the imaging process as a transferring of individual points (of various optical brightness) from the object plane to the image plane. Geometrical optics, taking into account magnification, *M*, dictates the new image point positions. The optical geometry of magnification will also diminish the brightness of each image point if $|M| > 1$, and, of course, increase the brightness of each image point if $|M| < 1$. When the object plane is relatively close to the lens the wave fronts *from each object point* will be distinctly spherical as they encounter the lens front surface, and since they must converge to a point in the nearby image plane, the wave fronts leaving the lens rear surface will be spherical also, not plane as in Fraunhofer phenomena detected at large distance *R*. So indeed, the so-called Fresnel diffraction equations will be in play here. Other people have contributed to the mathematical understanding of how wave fronts pass through optical systems and their names are also associated with various approaches, approximations, and forms of the equations which describe the nature of the propagating optical disturbance: Kirchoff, Rayleigh, Huygens, Sommerfeld. Goodman (1968) outlines this development in his Chapter 3.

The function which describes how an imaging system alters an object point as it transfers it from object plane to image plane can be called a spread function. With a magnification, $|M| = 1$, because of lens aberrations and the diffraction effects of apertures, an imaging system will typically create image points which are more spread, more broadened, than, an idealized object point. Even without aberration, diffraction will cause spreading. Though we are considering only aberration-free systems here, we will see in Chapter 9 that aberration effects can be "easily" incorporated into this linear systems approach. But in both kinds of imaging, coherent and incoherent, *as long as we are only interested in detecting intensity*, the spread functions in both imaging diagrams are obtainable from the Fraunhofer diffraction diagram (and process), even though the closeness of the imaging system object and image planes dictates a Fresnel diffraction treatment. The optical system, transferring a close object point into a close image point does so in a manner similar to Fraunhofer diffraction but now there will be additional phase contamination. With the exception of the additional phase contamination factors present in the Fresnel diffraction region, the Fraunhofer irradiance function (within a constant factor) is the spread function for imaging with incoherent light; the Fraunhofer amplitude function (within a constant factor) is the spread function for imaging with coherent light. *Since we image only with light intensity* the contaminating phase function factors, being complex functions, will multiply to unity (1).

6.1.4 Definition of Optical Transfer Functions

The optical transfer function for imaging with incoherent light is defined as the Fourier transform (or inverse Fourier transform) of the Fraunhofer irradiance function. The Fraunhofer irradiance, though purely real, can be asymmetric and hence the optical transfer function derived from it can be complex.

The optical transfer function for imaging with coherent light is defined as the Fourier transform (or inverse Fourier transform) of the Fraunhofer amplitude function. The Fraunhofer amplitude can be real or complex and hence the optical transfer function derived from it can be real or complex. A quick look at Figures 6.3 and 6.5, for coherent imaging and for Fraunhofer diffraction, will show that for coherent imaging the transfer function (the transform of the Fraunhofer amplitude function) will be mathematically proportional to the aperture function, that's nice; the aperture function is often real.

In both cases, being defined as Fourier transforms, the transfer functions represent the spatial frequency spectra of the point spread functions of the imaging system. This is the definition of the transfer functions. So, in a graph of a transfer function, the independent variable will be spatial frequency, a quantity such as lines per millimeter. The dependent variable, the transfer function itself, at each spatial frequency, describes the contrast, usually normalized to range from zero to one, zero being gray (no contrast), one being perfect black/white contrast. If an object grating of spatial frequency s_{xo} (pure white and black bars) is imaged by this system, the intensity contrast of adjacent bars in the image, at spatial frequency s_{xi}, is given by the transfer function. Perfect imagery of black/white things of frequency s_{xi} will have a transfer function value of 1.0. If in the image the bars are a complete washout, a continuous shade of gray, the transfer function value is zero; contrast at frequency s_{xi} is absent, even though light is present. Transfer functions are often

LINEAR SYSTEMS THEORY APPLIED TO OPTICAL SYSTEMS   139

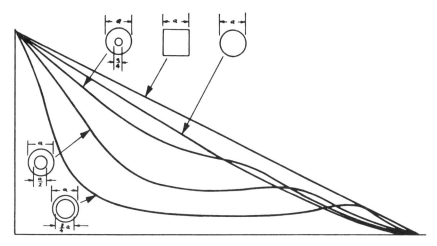

Figure 6.6 Cross-sections of transfer functions for several different apertures. Notice the one with a center obstruction of $a/2$; Figure 6.7 is its complete form. (E. B. Brown, *Modern Optics*, © 1965, Reinhold Publishing Corporation. (Reproduced with permission of Chapman & Hall.)

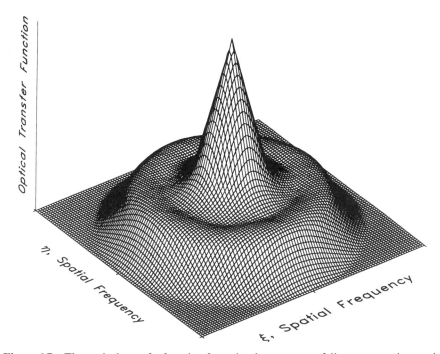

Figure 6.7 The optical transfer function for a circular aperture of diameter a and central circular obstruction of diameter $a/2$.

normalized to 1.0 at zero spatial frequency, and they decrease, eventually to zero, when the imaging system can no longer resolve bars at an increased spatial frequency; the contrast has gone to zero. Figure 6.6 is how they are typically portrayed for circular and symmetric functions. Figure 6.7 is the complete convolution of the aperture of Figure 6.6 with center obstruction of $a/2$. At most, perhaps only one-quarter of the figure is necessary; only one quadrant represents all positive spatial frequencies. Remember, the negative frequencies are mathematical artifacts, they are not real. Since Fourier created Fourier series (and series only *represent*, they are not physically *real* either), perhaps we should say that the negative frequencies are just one additional step away from reality, however necessary we find them to be.

Sometimes a transfer function can become negative, which means dark becomes light and light becomes dark, a black bar in the object appears to be imaged as white and vice versa, i.e., contrast reversal. See Figures 6.8 and 6.9.

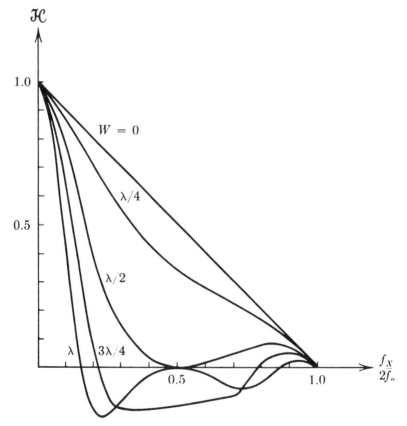

Figure 6.8 At the spatial frequencies where the transfer function is negative, those spatial frequencies are imaged with reversed contrast. (J. W. Goodman, *Introduction to Fourier Optics*, © 1968, reproduced with permission of the publisher, McGraw-Hill.)

LINEAR SYSTEMS THEORY APPLIED TO OPTICAL SYSTEMS 141

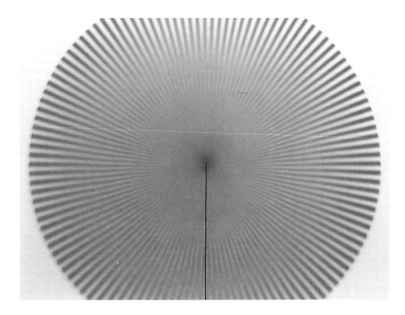

Figure 6.9 The low spatial frequencies have correct contrast. Starting on the line at the bottom of the figure, on approaching higher spatial frequencies the contrast (the resolution) goes to zero, but some resolution returns with negative (reversed) contrast in this example of a system with defocus only present in the upper and lower parts of the image. Three contrast reversals are visible in the original.

This reversal can occur if there are aberrations in the imaging system, such as defocus. This phenomena is not usually attributable to the real part of the aperture function, which is likely to be symmetric also, hence usually yielding an always positive transfer function. We will consider this again in Chapter 9, and there we will show other complete aperture convolutions for aperture functions which would also image with contrast reversal at some spatial frequencies.

6.1.5 Aperture Stop, Entrance and Exit Pupils

When a Fourier transform of a simple circular aperture is produced by Fraunhofer diffraction in the rear focal plane of a simple lens, the transform is that of the circle function. Often this aperture is defined by the circular rim of the lens or of the lens mounting ring. In a compound lens made of several lens elements of different size, and perhaps with apertures or diaphragms between lenses, there is one aperture the diameter of which "determines the amount of light reaching the image, this [physical aperture] is known as the *aperture stop*,"—Hecht (1987). The image of the aperture stop formed by all the lenses

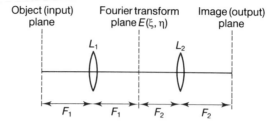

Figure 6.10 An imaging system in which there is also access to the Fourier transform of the object, probably a transparency.

which precede it is called the *entrance pupil*. The image of the aperture stop formed by all the lenses which follow it is called the *exit pupil*.

If an imaging lens with its aperture stop is used to produce a Fraunhofer diffraction pattern in the lens rear focal plane, the Fraunhofer *amplitude* so produced (the point spread function for coherent light imaging), is the Fourier transform of the optical disturbance at the exit pupil, (see Figure 6.10). The complex square of this, the Fraunhofer irradiance function, represents the point spread function for imaging with incoherent light.

In imaging systems the spread functions are obtained by Fraunhofer diffraction considerations of the wavefront at the exit pupil rather than directly from the aperture function. The transfer functions are obtained by Fourier transformation of the spread functions. We now know how to obtain all the functions present in the imaging diagrams. Aberrations in the system will also effect the wavefront at the exit pupil, and we will examine this in Chapter 9. Aberrations, though often nasty, need not become a limitation to our understanding.

6.2 IMAGING WITH INCOHERENT ILLUMINATION

There are two routes to the image formed with incoherent light. Let's examine an imaging system which uses incoherent and monochromatic illumination, linear in intensity (Figure 6.11). This kind of imaging is typified by conventional use of: 35 mm cameras, TV cameras, slide projectors, and imaging LCD or LED displays, etc., where the *objects* to be imaged radiate or are illuminated with incoherent light. Incandescent and fluorescent lamps produce incoherent illumination. Consider Figure 6.11.

Be careful of the following rather strange definition: the function $O(-x_i/M, -y_i/M)$ represents the geometrically magnified *perfect* image intensity. The actual image intensity, including the spreading effects of diffraction, is the convolution of the object function with the point spread function. Due to magnification the image intensity is reduced by the factor $1/M^2$. Read to the end of Section 6.2.1 and see if the definition is sensible.

IMAGING WITH INCOHERENT ILLUMINATION

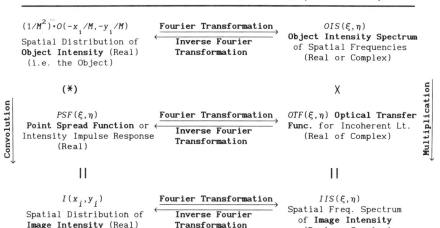

Figure 6.11 Imaging with incoherent light, |magnification| = M; see text for complete description.

From the diagram we see that we have two routes from the object to the image. The object, say a photo transparency, incoherently illuminated, can be thought of in two ways: (i) either as a two-dimensional array of periodic structures of various spatial frequencies, intensities, and orientations, or (ii) as a two-dimensional array of points of varying brightness.

6.2.1 Image Formation with Incoherent Light via the *OTF*, Using the Spatial Frequency Concept

A simple example of (i) would be an incoherently illuminated object function which is nothing more than a grating of white and black bars of identical width and spacing. Figure 6.12 (Washer and Gardner 1953) shows such an object grating; one with many grating spatial frequencies. These are exactly analogous to the one-dimensional square wave function, always positive; the spectrum of that function, fundamental frequency and harmonics, are well known and obtained with relative ease by Fourier analysis.

But the object of an optical imaging system is usually two dimensional; e.g., gratings in x and gratings in y. The frequencies of the bars are spatial frequencies, hence we could refer to the spatial frequency in the x- or y-directions or some intermediate direction. Our route in this example means calculating the two-dimensional Fourier transform (spectrum) of the object (the grating). But we know now that a simple lens is capable of performing the Fourier transform operation on a two-dimensional spatial scene (object). Thus,

144   LINEAR SYSTEMS THEORY APPLIED TO OPTICAL IMAGING

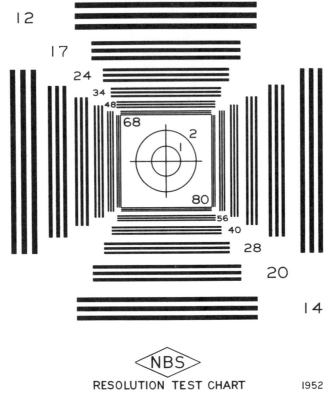

Figure 6.12 A bar target with high contrast and various spatial frequencies. (From F. E. Washer and I. C. Gardner, *Method for Determining the Resolving Power of Photographic Lenses*, National Bureau of Standards Circular 533, U.S. Government Printing Office, May 20, 1953.)

in Figure 6.11, for imaging with incoherent illumination of the object, one route from object to image consists of three steps:

1. Determination of the spatial frequency spectrum of the spatial distribution of the (always positive) intensity in the magnified perfect object by Fourier transformation. This we can do.
2. Multiplication of that (perfect object) spatial frequency spectrum by the incoherent optical transfer function (the *OTF* for our imaging lens) to produce the spatial frequency spectrum of the image intensity distribution in space. (We know how to get the *OTF* for imaging with incoherent light.)
3. An inverse Fourier transformation of that product specifies the spatial distribution of the always positive image intensity in space, i.e., the image; reduced in intensity by the factor $1/M^2$.

6.2.2 Image Formation with Incoherent Light Via the *PSF*, Using the Impulse Response Concept

The second route from object to image for a system with incoherent illumination involves the use of the incoherent light point spread function (*PSF*), or intensity impulse response, or the Fraunhofer irradiance function of the particular imaging system. We know how to get that, and after Chapter 9, we can do it even if aberrations are present. Treating the object as a two-dimensional array of points of varying intensity, the incoherent light point spread function (intensity impulse response) essentially determines the manner whereby each point of object intensity is changed to a point of image intensity. *Since we are detecting only intensity* the imaging system will convert each ideal object point into a Fraunhofer irradiance pattern which is characteristic of the exit pupil of the optical system. These image irradiance pattern points will overlap each other, thus decreasing resolution of the true points. The convolution operation is that which conceptually performs a point for point translation, from object plane to image plane, taking into consideration magnification, and diffraction and aberration spreading by the system's finite exit pupil. That is the second route from object to image; the convolution of the object intensity function with the *PSF* yields the image intensity function.

The use of the point spread function is at the heart of the definition of linear systems. If we know the system response to a single point, then we conceptually decompose the object into points, process each point, and finally form a superposition of all points placing them where the optical magnification geometry dictates; this becomes the image. The additional concept of invariance, and here we are only concerned with spatial invariance, stipulates that as we consider object points at various locations in the object plane these will be translated into image points at proper locations in the image plane, obeying the geometrical rules for inversion and magnification, but the functional form of the image points will not alter as locations are changed. Such a space-invariant system is also termed *isoplanatic*. If we restrict ourselves to isoplanatic regions then such spatial invariance concepts allow the superposition of points to be expressed as a convolution of the object function with the *appropriate* point spread function of the system, which is dictated by the system's exit pupil. We will illustrate this operation and others with examples in later chapters. Our theory is really *invariant* linear systems theory.

So far we have only considered aberration-free imaging systems. Aberrations can produce contaminating phase and intensity variations in an image; we haven't encountered this yet. The lenses will produce phase curvature as already discussed. For now, the wave fronts emerging from the exit pupils will be real and symmetric. Hence the point spread functions, derived from the shape of the exit pupil by Fourier transformation, should be relatively easy to determine.

From the simple examples we have been using one might get the impression that the incoherent *OTF* would always be mathematically real. After all, it is obtained by Fourier transformation from the always real *PSF*; the only *PSF*'s we have mentioned have been derived from real and symmetric aperture functions. But recall when we earlier (Chapter 5) Fourier transformed a *shifted*, and therefore *nonsymmetric* function, the resulting transform was complex; it contained a phase shift. Therefore if the *PSF* is *asymmetric*, its (*inverse*) Fourier transform will very likely contain a phase function factor. If the *PSF* is asymmetric, so will be the Fraunhofer amplitude; and how could it become asymmetric? We will find in Chapter 9 that a phase variation across the aperture can space shift the diffraction pattern into asymmetry. So a phase function in the aperture, by Fourier transformation, can create both an asymmetric Fraunhofer amplitude and intensity (*PSF*); and the asymmetry through inverse Fourier transformation creates a phase variation in the incoherent *OTF*. Many aberrations in an optical system are identical to generally undesirable phase variations (functions) caused by the lenses. It is possible to incorporate these phase variations (aberrations) into the exit pupil, and thence into the spread functions and transfer functions. We will see how this is done in Chapter 9. Additionally, the concept of apertures or exit pupils with phase functions *purposely designed into them*, would provide one more additional tool for optical processing. And then, why not apertures with time-varying amplitude and/or phase?

Thus, with incoherent illumination, we have our two routes to image formation using the concepts of invariant linear systems, the first utilizing spatial frequencies and Fourier transformation, the second utilizing the point spread function and linear superposition via the convolution operation.

6.3 IMAGING WITH COHERENT ILLUMINATION

Coherent imaging systems are not linear in intensity, but linear in the amplitude of light. The amplitude of each point in a coherent wave front bears a specific temporal and spatial relationship to every other point in the wave front which emanates from the coherently illuminated object. Hence, certain interference conditions are possible as the wave front moves through the optical imaging system. With coherent wave fronts, interference is determined by summations of light wave amplitudes, hence such a system is linear in light amplitude. Coherent imaging occurs in: x-ray crystallography; in imaging systems where illumination is by laser; and, under certain conditions, in microscopy.

An imaging system with coherent illumination would be diagrammed as in Figure 6.13.

Be careful of the following rather strange definition: the function, $U_0(-x_i/M, -y_i/M)$ represents the geometrically magnified *perfect* image amplitude and phase. The actual image amplitude, including the spreading effects of diffraction is the convolution of U_0 with the amplitude spread function. Due to magnification the image amplitude is reduced by the factor $1/M$. In such a system the image intensity, shown in Figure 6.13, is the product of the image

IMAGING WITH COHERENT ILLUMINATION 147

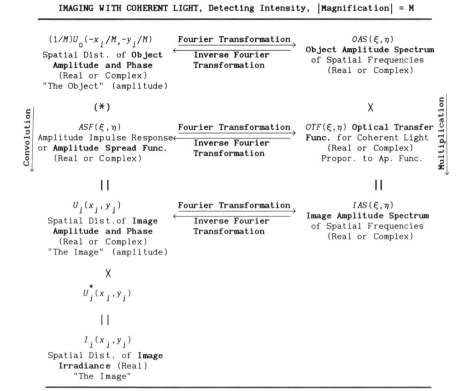

Figure 6.13 Imaging with coherent light, |magnification| = M; see text for complete description.

amplitude function with its complex conjugate. Thus the measureable quantity, the image intensity, will be reduced by $(1/M)^2$.

Likewise, the *observable* spectra, called power spectra, are the products of the *amplitude* spatial frequency spectra and their complex conjugates.

The impulse response above is identical in form to the Fraunhofer *amplitude*. Since the present system is linear in amplitude, the image amplitude function is formed by the convolution of the object amplitude function with an amplitude impulse response. This response function, as before, is obtained by Fourier transformation of the coherent optical transfer function. Since this amplitude response is identical in form to the Fraunhofer amplitude, the coherent OTF, by Fourier transformation, must be simply proportional to the aperture function, or exit pupil, of the imaging system, though of different dimensions. Indeed, under conditions of symmetry and normalization, they can be taken as of identical form. That should be very convenient, whence in an aberration-free system, the aperture is likely to be real and symmetric.

In actual computational use many of the functions mentioned in all the preceding pages are normalized to a common value, one, at the origin of their coordinate system. Also, certain operations mentioned previously are not reversible, notably the process of multiplying complex functions, wherein phase information is lost, seemingly irretrievably lost. Thus the intensity or power functions are not necessarily uniquely related to their amplitude factors, though the amplitude functions are generally unique and of special interest here. The phase recovery problem continues to be an area of active interest and research. A recent paper by Pérez-Ilazarbe (1992) includes a lengthy list of references.

In recent years coherent systems have seen expanded use, not necessarily for imaging, but in systems generally termed coherent-optical processors, -optical correlators, or -optical computers, where the input function may derive from opto-electronic or photonic means, rather than purely optical means, and may be continuously variable in time.

6.4 TRANSFER FUNCTIONS BY CONVOLUTION; CONCEPTUAL BASIS

In Figure 6.5, the diagram for Fraunhofer diffraction (reproduced here as Figure 6.14), we see that a convolution operation obtains for us the optical transfer function for imaging with incoherent light. The convolution operation is one

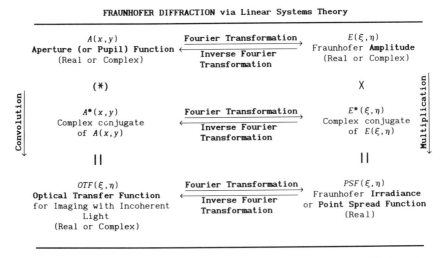

Figure 6.14 Fraunhofer diffraction via linear systems theory. One could also refer to this diagram as the convolution theorem applied to Fraunhofer diffraction. The convolution operation always involves two functions; in these optical applications it is sometimes a self-convolution (folding included) of apertures with themselves; but also, as in Figure 6.1, where an input function is convolved with an impulse response function.

TRANSFER FUNCTIONS BY CONVOLUTION; CONCEPTUAL BASIS 149

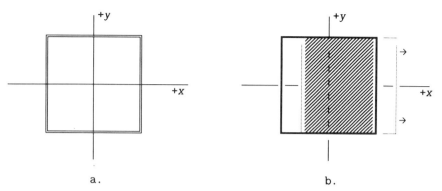

Figure 6.15 In (a), a second fictitious identical aperture has been folded over atop the real one. In (b), the top one is shifted (slid) positively while keeping track of the amount of overlap area as a function of the amount of shift.

with which the reader may not have familiarity. We will describe one of the simplest—the convolution of a square aperture with itself. This is a geometrical conceptualization.

Given the square aperture:

1. Create a second fictitious but identical square aperture.
2. With the two side by side, fold the fictitious square over onto the top of the real one. This folding is called *faltung* in German.
3. Calculate the overlap area of the two squares. For squares both of which have an area of one (1.0), the overlap area is now 1.0, see Figure 6.15(a).
4. Now shift (slide) the fictitious square in the $+x$-direction, and as a function of the amount of shift, keep track of the overlap area (Figure 6.15(b)). It is easy to see that the rectangular overlap area of narrowing width, shifted $+x$, goes linearly to zero. Do the same for a shift in the $-x$-direction. Plotting area as a function of shift should now appear like a triangle standing on the x-axis (Figure 6.16).
5. Repeat the above but for shifts in the $+y$ and $-y$ directions. Figure 6.17 illustrates the result.
6. Instead of doing all x shifts first, then y shifts, if we had done some x, then some y, more x, more y, etc., the following pattern (Figure 6.18) would have evolved. The whole thing could have been created if we shifted *both* apertures *diagonally* at the same time.

The four plots in Figure 6.19 illustrate the formation of Figure 6.18 as a function of the shifts. The horizontal plane represents the axes of shifts while the z-axis represents the overlap area as a function of shifts in x and y. It is only along slices parallel to the x- or y-axes that this function decreases linearly.

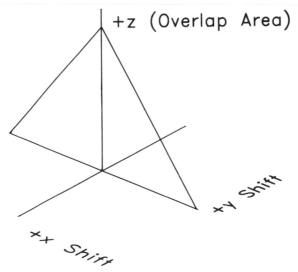

Figure 6.16 The operation of Figure 6.15 is repeated for negative shift; the overlap area(s) for $\pm x$ shifts go linearly to zero.

The surfaces, however, are warped; and perfectly *real apertures* or exit pupil functions will never have negative values ($z < 0$) in their transfer functions. The axes in the horizontal plane will need to be scaled (converted) to spatial frequencies; that is what they represent in a transfer function. But fundamentally, this is how one does a convolution. For simple and real apertures

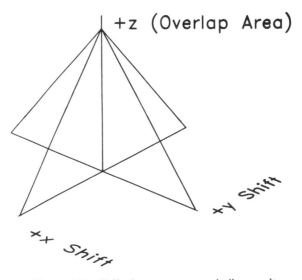

Figure 6.17 Shifts in $\pm y$ generate similar results.

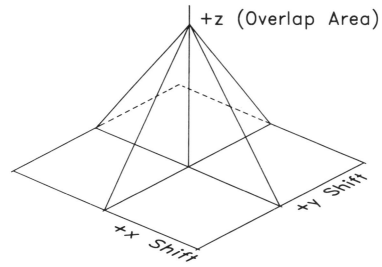

Figure 6.18 Had we mixed the shifts, some x, them some y, repeatedly, this results. Thus the complete convolution function will be proportional to the warped surface on this "tent" frame and its x–y-plane square boundary. See Figures 6.19(a)–(d).

a mathematical expression can be derived for the overlap area as a function of shift, e.g., $z = f$(shift in x, shift in y). Though it is no more than a geometrical calculation, it can easily become a bit difficult. It is a job nicely suited for computers.

It is interesting to compare the shape of the above convolution with the convolution of a circular aperture (or circle function) with itself. Refer to all parts of Figure 6.20, where we see the representation of a circular aperture, (Figure 6.20(a)); two shifted apertures wherein the overlap area, for this amount of shift, is represented by the 2nd (the top) floor (Figure 6.20(b)); and four plots, Figures 6.20(c) to 6.20(f), showing progressive creation of the circular aperture's transfer function for imaging with incoherent light. Goodman (1968) and others have derived the expression for this function. Using Goodman's notation,

$$\mathcal{H}(\rho) = (2/\pi)[\cos^{-1}(\rho/2\rho_0) - (\rho/2\rho_0)\sqrt{1 - (\rho/2\rho_0)^2}], \quad \text{for } \rho \leq 2\rho_0$$
$$\mathcal{H}(\rho) = 0, \quad \text{otherwise}$$
(6.1)

The function $\mathcal{H}(\rho)$ is what we have called the OTF; it is real here. ρ is the dependent variable spatial frequency, equal to our $\sqrt{\xi^2 + \eta^2}$, while $2\rho_0$ is the cutoff spatial frequency for imaging with incoherent light.

$$2\rho_0 = \frac{\text{aperture diameter}}{\lambda \cdot (\text{image distance})} \quad (6.2)$$

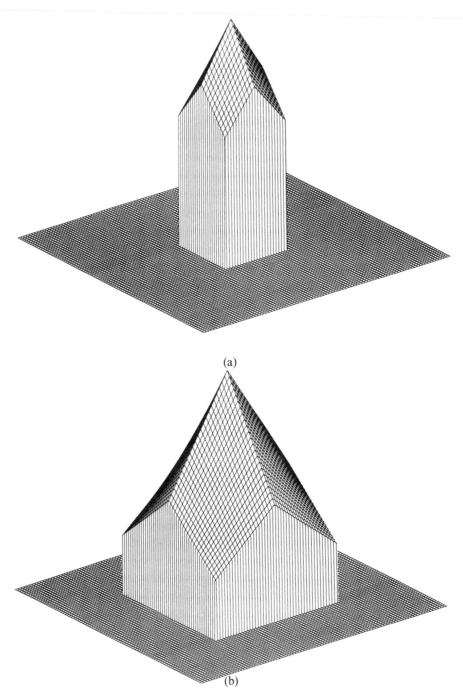

Figure 6.19(a)–(d): The computer generates in steps of shift the surface representing the convolution of the square function. The surfaces are warped except in directions parallel to the axes.

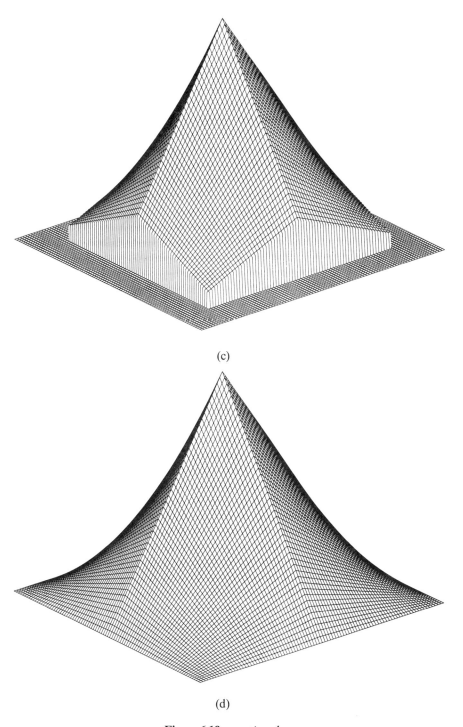

(c)

(d)

Figure 6.19—*continued.*

The quantity $2\rho_0$, equivalent to what we would call a cutoff ξ or η, is that maximum spatial frequency beyond which there is no resolution, no contrast, for incoherent imaging.

The quantity ρ_0 will be the cutoff spatial frequency for coherent imaging. This fact follows from the definition of the *OTF* for coherent imaging being proportional to the aperture function itself; no convolution needed. This means that if we have determined the cutoff frequency for imaging with incoherent light, $2\rho_0$, then if we construct the proper *OTF* for imaging with coherent light (proportional to the aperture or exit pupil function) it can be rescaled to indicate the cutoff frequency at ρ_0. This does not necessarily mean that better (twice better) resolution can be obtained with an incoherent imaging system. Check the references on this point.

A word about resolution is appropriate here. Goodman is quite detailed about its determination; let's try something easier. In most general physics textbooks there is a discussion about resolution in telescopes. The criteria used is that of Rayleigh which can be phrased as the following: with a telescope imaging two equally bright stars which are separated by a black space and surrounded by blackness (two bright bars separated by a black bar), the two stars are *just resolvable* if in the telescope focal plane (the Fourier transform plane for the aperture) the central maximum of the Fraunhofer irradiance of one star coincides with the first minimum of irradiance of the other star. When this occurs, the angular separation of the two stars, and hence the angular separation of the two irradiance functions (vertex assumed at the telescope lens on the optical axis), is given by

$$\Delta\theta \cong \sin\Delta\theta = \frac{1.22\lambda}{\text{aperture diameter}} \quad \text{for a circular aperture.}$$

$$\Delta\theta \cong \sin\Delta\theta = \frac{\lambda}{\text{aperture diameter}} \quad \text{for a square aperture.}$$

(This last $\Delta\theta$ you may recall from a general physics course as the "angular resolution" of a square aperture. The $\Delta\theta$ for a circular aperture is usually mentioned also in a first physics course. We derive it in Section 7.2.3.) Thus, always using the small angle approximations, the equations above mean that the space separation at the Fourier transform plane (the focal plane), Δl, would be provided by the statement, $\Delta\theta \cong \tan\Delta\theta = \Delta l/F$. Hence this space separation is $\Delta l = F\Delta\theta$; we would call it the image separation at *just resolved*, probably measured in millimeters. The spatial frequency just resolved $\xi(\text{maximum}) = 1/\Delta l = 1/F\Delta\theta$ or

$$\xi(\text{maximum}) = \frac{\text{aperture diameter}}{\lambda F} \quad \text{for a square aperture}$$

This is almost Goodman's $2\rho_0$, our (6.2). In a general imaging system, the image is formed probably not in a rear focal plane (not at F), but at an image distance, d_i. The smallest angle of resolution approximations can still be made but $\Delta\theta$ will now be $\Delta\theta \cong \tan \Delta\theta = \Delta l/d_i$, rather than $\Delta l/F$ as above. The minimum spatial separation between resolvable points, $\Delta l = d_i \Delta\theta = d_i \lambda/$(aperture diameter), for a square aperture. The maximum resolvable spatial frequency $\xi(\text{maximum}) = 1/\Delta l$,

$$\xi(\text{maximum}) = \frac{\text{aperture diameter}}{\lambda d_i}$$

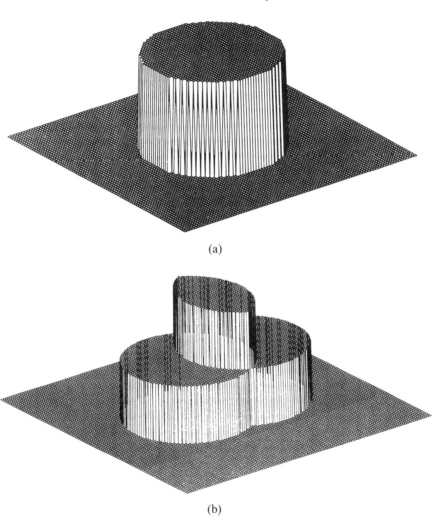

(a)

(b)

Figure 6.20 (a) A circular aperture representation. (b) Two circular apertures (one folded), shifted. Note the amount of overlap area. Can you calculate the amount of overlap area?

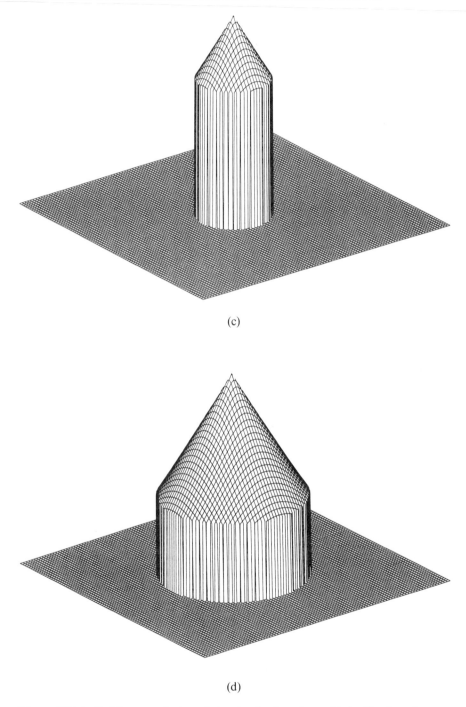

(c)

(d)

Figure 6.20(c)–(f): The convolution of the circular function, with itself, takes shape as the $\pm x$ and $\pm y$ shifts take place. The surface is warped everywhere.

TRANSFER FUNCTIONS BY CONVOLUTION; CONCEPTUAL BASIS 157

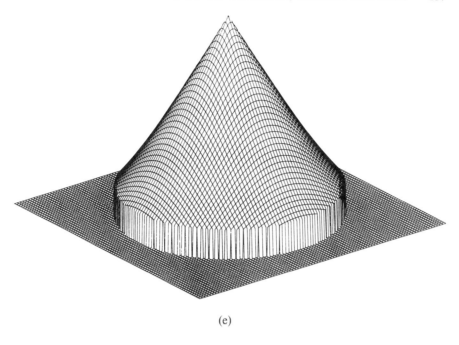

(e)

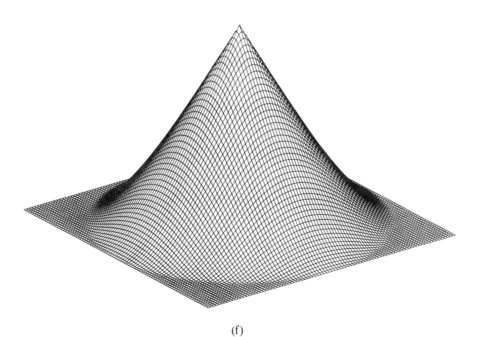

(f)

Figure 6.20—*continued*

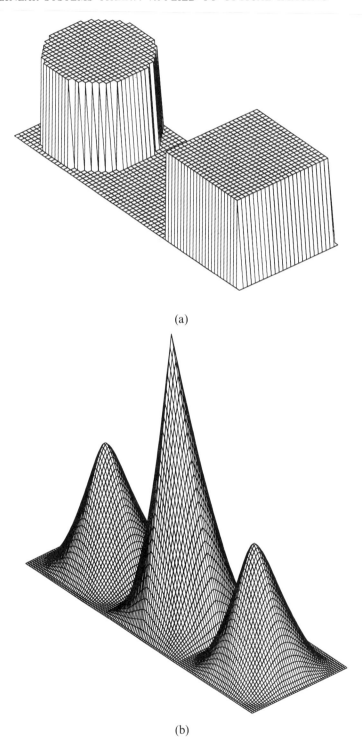

(a)

(b)

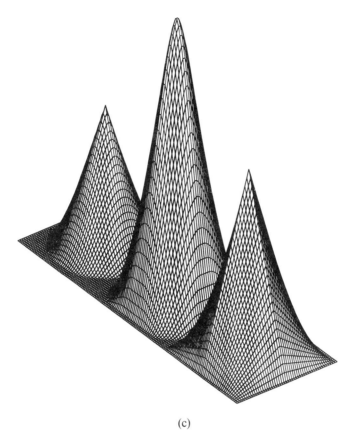

(c)

Figure 6.21 (a) Circular and square functions, and a spacing, all the same width. (b) If a second aperture is layed on *without folding*, and shifted, this shape results; see text. (c) If the second aperture is folded before laying on, and then shifted, this surface results; *faltung* is important.

For incoherent imaging this is the same cutoff frequency obtained by Goodman for the square aperture; however his function, (6.1), obtained by mathematical convolution, also describes the lessening contrast as a function of spatial frequency. Convolution, yielding the *OTF*, is a useful tool.

Convolutions of symmetric aperture functions will obviously yield symmetric *OTF*'s. In a convolution, we can illustrate the importance of the folding operation which is done before the shifting. Figure 6.21(a) is a representation of a double aperture, a circle and square both of the same width, and separated by the same amount. Think of the fictitious aperture, overlayed *without folding*, and shifted to the right. We start at maximum overlap area, at the central peak of this operation. The first third of the entire shift to the right (i.e., a shift equivalent to one width in length) moves the circle off the circle, and the square off the square to where there is no overlap: the overlap area has gone to zero. The next third of the entire rightward shift moves the *circle onto the square*, increasing overlap area; the last third of the total shift moves the *circle off the square* in a symmetrical manner. Shift to the left, the left side of the result will look the same since circle and square overlap again. This is represented in Figure 6.21(b).

But now, fold the fictitious aperture first so that we begin the process not with square on square and circle on circle, but with two pairs of "circle on square" overlaps; our initial value here, the central peak of the convolution, will be less than that for Figure 6.21(b). In our earlier convolution of the single square aperture (Figure 6.19(d)), and single circular aperture (Figure 6.20(f)), both results had sharp peaks at the origin. Notice in Figure 6.21(c) the central peak is rounded. The first third of the total shift to the right moves the overlap area from central peak to zero; and then, in the final two-thirds of the shift to the right, we have a square moving onto and then off another square. The result should show on the right side a function that looks like the square convolution alone. Shift to the left and the last two-thirds of the shift is that for the convolution of a single circle with itself; readily apparent on the left side. The folding, *faltung*, is important.

6.5 CALCULATION EXAMPLES

Here is the calculation for a specific aperture consisting of two open squares, of side $2a$, separated by $2a$, in an otherwise opaque screen. The value of the aperture function $A(x, y)$ is unity, i.e., $A = 1.0$; its shape is defined by the integration limits (Figure 6.22).

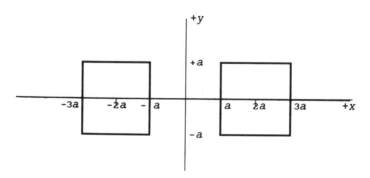

Figure 6.22 Two square apertures of side $2a$, spaced $2a$.

6.5.1 Amplitude Impulse Response

Aperture Function $\xrightarrow{\text{Fourier Transformation}}$ Amplitude Impulse Response (Fraunhofer Amplitude)
$A(x, y)$ $E(u, v)$

$$E(u, v) = \int_{-\infty}^{+\infty} \int_{-\infty}^{+\infty} A(x, y) e^{i(ux+vy)} \, dx \, dy$$

$A(x, y)$ being *real* and *even* allows us to write

$$E(u, v) = 2 \int_{a}^{3a} 1 \cdot \cos(ux) \, dx \cdot 2 \int_{0}^{a} 1 \cdot \cos(vy) \, dy$$

Carrying out the integration and simplifying,

$$E(u, v) = 8 \cdot \frac{\sin(au)}{u} \cdot \frac{\sin(av)}{v} \cdot \cos(2au)$$

Multiplying the right side by a^2/a^2,

$$E(u, v) = 8a^2 \cdot \frac{\sin(au)}{au} \cdot \frac{\sin(av)}{av} \cdot \cos(2au)$$

Consolidating the constant factors into C and using abbreviated notation,

$$E(u, v) = C \cdot \text{sinc}(au) \cdot \text{sinc}(av) \cdot \cos(2au)$$

For this aperture this is the amplitude impulse response or Fraunhofer amplitude. It is also the form of the spread function for imaging with coherent light.

6.5.2 Point Spread Function

The point spread function is the form of the square of $E(u, v)$.

$$E^2(u, v) = C^2 \cdot \text{sinc}^2(au) \cdot \text{sinc}^2(av) \cdot \cos^2(2au)$$

This is the form of the intensity impulse response, the point spread function for incoherent imaging, or the Fraunhofer irradiance.

6.5.3 Transfer Function for Imaging with Incoherent Light

For imaging with *incoherent* illumination, the incoherent *OTF* can be determined, in this case, either by calculating the inverse Fourier transform of E^2, the point spread function, or by convolving the aperture function itself. Either is probably best done by computer. For this simple example we choose the latter approach, convolution. Some steps can be taken beforehand to yield a normalized transfer function directly.

Normalize the aperture for the wavelength of light being used, and for image distance, d_i, as we have done in Figure 6.23. The *normalized OTF* (incoherent) will be the convolution of this double aperture with itself *divided* by the total overlap area before shifting the double aperture over itself, i.e.,

$$OTF_i(\xi', \eta') = (1/N) \int_{-\infty}^{+\infty} A(x', y') \cdot A^*(x' - \xi', y' - \eta') \, dx' \, dy'$$

A further simplification is to call the shifts of the normalized aperture, ξ' and η'. Geometrical arguments may be used in this evaluation and separate shifts, ξ', and then η', can be made. Because of the symmetry we choose to make both shifts at the same time and instead of shifting one double aperture ξ' and η' over the other, we will shift one double aperture $+\xi'/2$ and $+\eta'/2$ and the other $-\xi'/2$ and $-\eta'/2$. See Figure 6.24.

$$OTF_i(\xi', \eta') = (1/N) \int_{-\infty}^{+\infty} A(x' + \xi'/2, y' + \eta'/2) \cdot A^*(x' - \xi'/2, y' - \eta'/2) \, dx' \, dy'$$

The convolution area desired is twice one of the cross-hatched areas. It equals

$$[(3a/\lambda d_i - \xi'/2) - (a/\lambda d_i + \xi'/2)][(a/\lambda d_i - \eta'/2) - (-a/\lambda d_i + \eta'/2)]$$
$$= 2(2a/\lambda d_i - \xi')(2a/\lambda d_i - \eta')$$

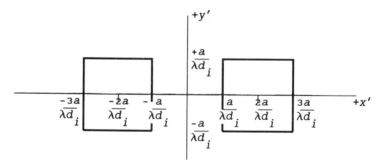

Figure 6.23 Normalizing Figure 6.22, for light wavelength, λ, and image distance d_i.

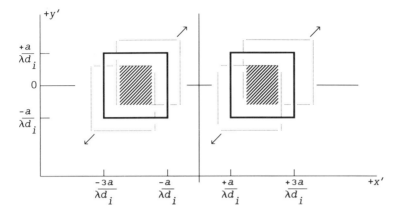

Figure 6.24 Second aperture layed on after folding, and all shifts, $\pm x'$, $\pm y'$, undertaken at the same time; see text. Caution: only try this when accompanied by an adult.

Overlap before shifting, $N = 2(2a/\lambda d_i)(2a/\lambda d_i) = 8a^2/\lambda^2 d_i^2$. The reciprocal of this will normalize the function to unity at the origin. The normalized first part of the calculation is:

$$OTF_i(\xi', \eta') = (\lambda^2 d_i^2/4a^2)(2a/\lambda d_i - \xi')(2a/\lambda d_i - \eta')$$
$$= (\lambda d_i/2a)(2a/\lambda d_i - \xi') \cdot (\lambda d_i/2a)(2a/\lambda d_i - \eta')$$
$$OTF_i(\xi', \eta') = (1 - (\lambda d_i/2a)\xi') \cdot (1 - (\lambda d_i/2a)\eta')$$

This is only the central part of the $OTF_i(\xi', \eta')$. The next step is to simply remove the primes from ξ and η.

$$OTF_i(\xi, \eta) = [1 - (\lambda d_i/2a)\xi] \cdot [1 - (\lambda d_i/2a)\eta] \qquad \text{(central part)}$$

Clearly, this is only a geometrical calculation; the reason the result comes out so swiftly in proper dimensions is because of the manner in which we normalized the aperture prior to the convolution. As the shift continues, one square will overlap the other square; a similar calculation yields a peak only half the height of the central one and the final ξ cutoff frequency at $6a/\lambda d_i$; η cutoff at $2a/\lambda d_i$. The $OTF_i(\xi, \eta)$ would be similar to that shown in Figure 6.25(b) which is the convolution of two rectangle functions (Figure 6.25(a)), with a central spacing the same as their widths. Often the cutoff frequency is stated for unit image distance, $d_i = 1$; the cutoff frequencies for the above example are $6a/\lambda$ and $2a/\lambda$. A computer is useful here.

In Figure 6.26(a) can you understand why this convolution of the two rectangles of Figure 6.26(b) has two flat, zero regions? And lastly, Figure 6.27(b)

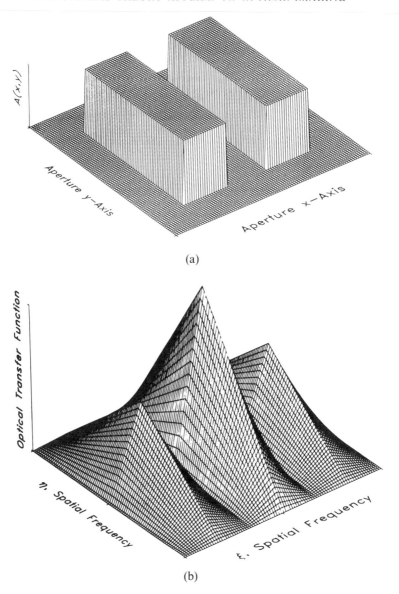

Figure 6.25 (a) Two rectangle functions (slits), of width equal to their spacing. (b) The convolution of the rectangle function of 6.25(a) with itself.

shows the result of convolving two "diagonal squares," see Figure 6.27(a); does it seem correct?

Obviously, convolution calculations might be fun to do; it's nice analytic geometry, but it could become damn hard. Most, but not all, of the convolutions just seen were *calculated* by a computer; we did some by hand. All the detailed plots were carried out by a computer.

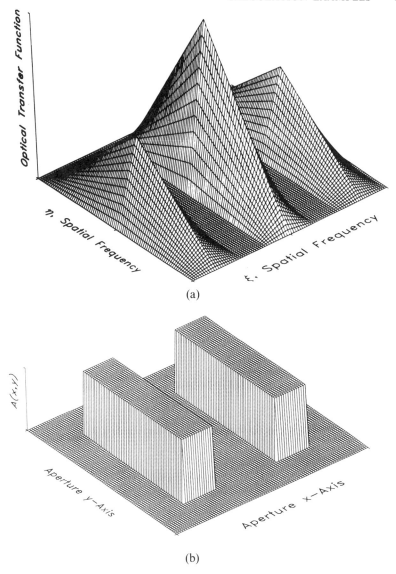

Figure 6.26 (a) is the convolution of the double rectangle function of 6.26(b) where the spacing is larger than the width.

6.5.4 Transfer Function for Imaging with Coherent Light

For imaging with *coherent* illumination, the coherent optical transfer function, $OTF_c(\xi, \eta)$ is proportional to the aperture function, and of different dimensions. A normalized expression can be *written* for the aperture function, though it is not too useful in programming for plotting (at least my PC software doesn't understand it).

$$A(x, y) = \text{rect}\left(\frac{x - 2a}{2a}, \frac{y - 0}{2a}\right) + \text{rect}\left(\frac{x + 2a}{2a}, \frac{y - 0}{2a}\right)$$

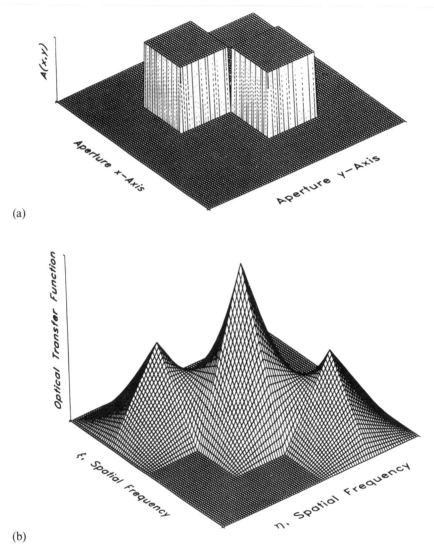

(a)

(b)

Figure 6.27 (a) How would you like to convolve this "diagonal squares" function with itself? See Figure 6.27(b) where a computer calculated and plotted self-convolution of the function in 6.27(a) is shown.

A rescaled plot of the aperture itself would be sufficient. The *OTF* would look something like Figure 6.25(a), with squares instead of rectangles, continuing our example of two square apertures. Rescaling the x,y axes over to ξ,η for the *OTF* with coherent light, the cutoff frequencies are each half those calculated for the incoherent case. The ξ cutoff is $(6a/\lambda d_i)/2$; the η cutoff is $(2a/\lambda d_i)/2$.

REFERENCES

On optical radiation quantities and units: Pinson (1985); Williams and Becklund (1972); O'Shea (1985).

On linear systems theory applied to optics: Goodman (1968); Banerjee and Poon (1991); Gaskill (1978); Papoulis (1968); Williams and Becklund (1989); Levi, Vol. 2 (1980); Saleh and Teich (1991); Reynolds, DeVelis, Parrent, and Thompson (1989); Livesley (1989); Blackledge (1989); Brown (1965); Carlson (1977).

On the phase problem: Pérez-Ilzarbe (1992); Leith and Upatnieks (1962); Greenway (1977).

FOR FURTHER CONSIDERATION

1. Try to sketch out, with pencil and paper and your imagination, the convolution of a **T**-shaped aperture. Shift to the right first and get that profile, then shift up, then down. You'll find a computer generated one later in this book, near the "T"-shaped aperture at the end of Section 10.5, though that one is for a "well-proportional **T**". Shall we call the one above a "square T"?

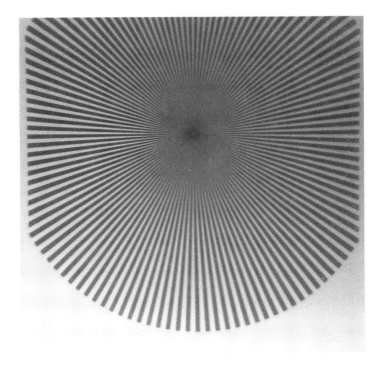

Figure 6.28 Image of radial bar target. The slit was |.

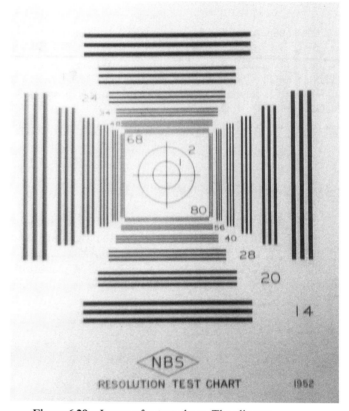

Figure 6.29 Image of a test chart. The slit was ———.

2. Figure 6.28 is a radial bar target imaged with incoherent light through a narrow slit aperture. Can you determine whether the slit was vertical or horizontal?

3. The test chart of Figure 6.29 was also imaged with incoherent light through a narrow slit aperture. Was the slit vertical or horizontal?

7

FOURIER OPTICAL TRANSFORMATIONS BY COMPUTER

It is now possible to explore the Fourier transform relationships which result from the scalar theory of diffraction, in 3-space, by personal computer. All aspects of Fraunhofer diffraction of essentially any two-dimensional aperture can be obtained rather simply: amplitude and phase, irradiance (point spread function), modulation and phase transfer functions; real or complex. Additionally some aspects of Fresnel diffraction may also be portrayed; we refer to this in Chapter 11.

The above statement is intended not only to introduce this chapter but to show contrast with past articles and books in which, theoretically or experimentally, diffraction is often treated as one-dimensional; or if two-dimensional, we are presented with only irradiance photographs. We can now have it all. Science journals have led the way in the use of 3-space computer graphics, but how they do it is only briefly described, if at all. We hope to remedy that here. Admittedly, in the matrix methods we will demonstrate, one may need ingenuity creating the aperture matrix. But once that is done, all other functions are created almost "automatically." We must also admit that it is difficult to keep pace with the many rapid advances in computer graphics technology.

7.1 PRIMITIVE APPROACHES

When we began this project in the late 1970's our approach was to calculate by hand the Fourier transforms required, $E(X, Y) =$ Fourier transform of $A(x, y)$, **and then ask late 1970's computers to draw the transform in 3-space, telling the computer to plot** $Z(X, Y) = E(X, Y)$.

170 FOURIER OPTICAL TRANSFORMATIONS BY COMPUTER

In those days, computer screen printout was usually unsatisfactory, unless one had expensive equipment; we did not. A good output device was a CalComp plotter, which required a good ball-point pen on a rotary drum. We didn't have that either. And there were very few people around us who understood the use of 3-space hidden line graphic algorithms. Our cost-efficient, but time-inefficient approach, was the use of a desktop computer with the "specification" of 61670 bytes of useable read/write memory, and a flat-bed pen plotter (Figure 7.1). That "61670 bytes" was the only piece of "computer-ese" specification. All other specifications were electrical and physical. The computer was very quick if we *did not* ask for a hidden line drawing.

For hidden line drawings, and with only 62k bytes, the procedure was to start drawing in the background, but before the next segment of a single line was drawn, the entire foreground pattern was calculated to see IF the next segment could indeed be drawn. Then move on to the next segment of the same line, and repeat the process. The plotting was done on-line with the computer; the software was written that way; there was insufficient memory to do our plots otherwise. Very seldom we requested a plot which required two weeks; and if there were no local violent thunderstorms we got results. In 1993 those same graphics take less than five minutes, on a 386 PC with a coprocessor and laser printer.

In the late 1980's we switched to 286 and then 386 personal computers with coprocessors and software packages, SURFER$^*_®$ and MATLAB$^*_®$. **Again, if one**

Figure 7.1 Author R. Wilson, pen plotter and 62 kbyte computer.

can solve by hand for the Fraunhofer diffraction amplitude, e.g., the

$$Z(X, Y) = (\text{constant})(ab)\,\text{sinc}(\pi aX)\,\text{sinc}(\pi bY)$$

for the $a \times b$ rectangular aperture, then SURFER can plot this $Z(X, Y)$ very easily at any orientation and with great resolution of detail. (Remember, in the transform plane the Cartesian coordinates are X, Y, Z, not x, y, z.) All of our earliest plots were drawn by a Hewlett Packard 9872B ($11'' \times 17''$ paper size) flat-bed pen plotter. At first, data for the plotter came from the 62 kbyte computer (using HP software), then later the data came from a PC (using SURFER) to the plotter via a Black Box® Corporation RS-232↔IEEE 488 interface converter. Our most recent plots are from a Hewlett-Packard LaserJet 4L printer at 300 dots per inch.

SURFER, in its Grid program, allows one to square the hand-calculated amplitude function and plot the irradiance or point spread function. If the Fraunhofer amplitude is mathematically complex, e.g., phase shifted, and one has calculated both the real (Re) and imaginary (Im) parts, then one can also have SURFER plot phase in the amplitude pattern, $\Phi(X, Y) = \arctan[\text{Im}(E)/\text{Re}(E)]$. SURFER also allows one to do element-by-element operations between two grids of X–Y data and hence create a third output grid of X–Y data for plotting or for further manipulation and simulation. For example, addition, subtraction, or multiplication of two grids is possible. SURFER is easy to use.

But SURFER cannot do all of the mathematical operations required in the Fourier approach to diffraction; it has limited math utilities for dealing with surfaces. The convolution theorem as mapped out in Figure 7.2 is a very

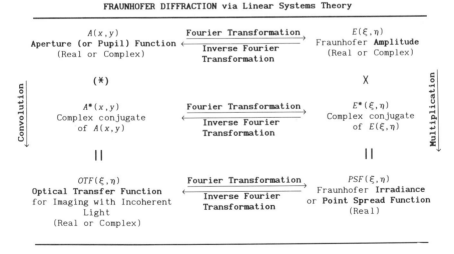

Figure 7.2 A Fourier description of Fraunhofer diffraction. The *OTF* requires normalization for expression in (ξ, η) coordinates.

convenient way to remember the Fourier approach. We mapped out this approach in Chapter 6 on Fourier theory. Examination of Figure 7.2 will show it to be a complete description of Fraunhofer diffraction, and more. It is the MATLAB software, and other such utilities, which allow one to do: two-dimensional fast Fourier transformations, inverse transformations, convolutions, correlations, etc., **on data in matrix form.**

7.2 SURFACE PLOTTING HAND-CALCULATED $Z(X, Y)$

Let's first examine the procedure whereby one calculates *by hand* **the Fourier transform, $E(X, Y)$, and then offers it to the computer for plotting** as $Z(X, Y)$. Either SURFER or MATLAB can plot our calculated function; SURFER's menus make it very easy.

7.2.1 The Square Aperture; Preparation for SURFER

In Chapter 5 we implied the calculation for the Fourier transforms for the square and rectangular apertures. Let's go through the steps here.

For the square aperture of dimensions a by a, the Fourier transform in terms of spatial frequencies is obtained from

$$E(\xi, \eta) = \int_{-a/2}^{+a/2} \int_{-a/2}^{+a/2} A(x, y) e^{-i2\pi(\xi x + \eta y)} \, dx \, dy \tag{7.1}$$

where $A(x, y)$ has a value of 1 inside the aperture, where x and $y \leq |a/2|$, and zero outside, where x and $y > |a/2|$. The integral is separable:

$$E(\xi, \eta) = \int_{-a/2}^{+a/2} e^{-i2\pi\xi x} \, dx \cdot \int_{-a/2}^{+a/2} e^{-i2\pi\eta y} \, dy$$

Since the aperture function is an even function in both the x- and y-directions we could have used the Fourier cosine transform. But continuing:

$$E(\xi, \eta) = \frac{1}{-i2\pi\xi} e^{-i2\pi\xi x} \Big|_{-a/2}^{+a/2} \cdot \frac{1}{-i2\pi\eta} e^{-i2\pi\eta y} \Big|_{-a/2}^{+a/2}$$

$$= \frac{1}{-i2\pi\xi} (e^{-i2\pi\xi a/2} - e^{+i2\pi\xi a/2}) \cdot \frac{1}{-i2\pi\eta} (e^{-i2\pi\eta a/2} - e^{+i2\pi\eta a/2})$$

$$= \frac{1}{\pi\xi} \left(\frac{e^{i2\pi\xi a/2} - e^{-i2\pi\xi a/2}}{2i} \right) \cdot \frac{1}{\pi\eta} \left(\frac{e^{i2\pi\eta a/2} - e^{-i2\pi\eta a/2}}{2i} \right)$$

SURFACE PLOTTING HAND-CALCULATED Z(X, Y)   173

We have moved the $-2i$ under the exponentials and used the minus sign to switch the order of the exponential terms. You probably recognize them then as $\sin(\pi\xi a)$ and $\sin(\pi\eta a)$. So we write

$$E(\xi, \eta) = (1/\pi\xi) \sin(\pi\xi a) \cdot (1/\pi\eta) \sin(\pi\eta a)$$
$$E(\xi, \eta) = (a/\pi\xi a) \sin(\pi\xi a) \cdot (a/\pi\eta a) \sin(\pi\eta a)$$
$$E(\xi, \eta) = a^2 \operatorname{sinc}(\pi a\xi) \operatorname{sinc}(\pi a\eta)$$

We have used our definition of

$$\operatorname{sinc}(X) = \frac{\sin(X)}{X}$$

This identical result is obtained if a positive exponential is used in (7.1).

Now the computer doesn't know beans about ξ and η, all it understands is x, y, and z for 3-space plotting. You want the sinc functions to go out far enough in ξ and η, the spatial frequencies, to show a reasonable number of maxima and minima. So if you let $a = 1$, for convenience, and let ξ and η, i.e., x and y, go to maximum values of ± 5, then a sinc function, equal to, say, $\sin(\pi x)/\pi x$, will go out to $\pm 5\pi$ radians, ± 15.7 radians (± 900 degrees if you prefer). Even though $\sin(\pi x)$ has three maxima at $x = +\frac{1}{2}$, $+\frac{5}{2}$, and $+\frac{9}{2}$, in this positive range, $\sin(\pi x)/\pi x$ will have but two, and one at the origin. Figure 7.3 contains a SURFER generated plot in 2-space of $\sin(\pi x)/\pi x$.

So going to the computer, if we want correct axis labeling, we ask for a plot of $z = (\sin(\pi x)/\pi x) \cdot (\sin(\pi y)/\pi y)$, it is being understood that $a = 1$, and plot

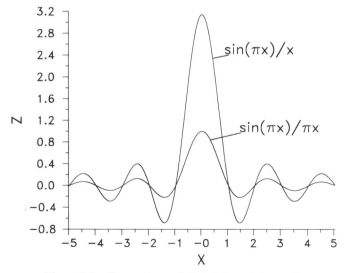

Figure 7.3 Comparison of $\sin(\pi x)/x$ and $\sin(\pi x)/\pi x$.

174 FOURIER OPTICAL TRANSFORMATIONS BY COMPUTER

over ranges of x and y, ± 5. (We need to know that the computer reads the (πx) and (πy) in radians and not degrees.)

The π's in the denominator could be omitted; in Chapter 5 we determined that there was already a constant outside the integral with which we began. However, at the origin of coordinates, at $(0, 0)$, $\sin(x)/x$ and $\sin(y)/y$ both have a value of unity (one). **More often than not these optical functions, regardless of their actual value at $(0, 0)$, are normalized such that their value at $(0, 0)$ is one.** So, for correct labeling purposes, if (πx) and (πy) are the sine arguments, then πx and πy had both better be in the denominator. Consult Figure 7.3 again; the function $\sin(\pi x)/x$, also contained there, has a value of about 3.2 at $(0, 0)$.

So here is the SURFER plot of our calculated Fourier transform, $E(\xi, \eta)$, of a square aperture function, $A(x, y)$, with uniform illumination, Figure 7.4, with $a = 1$. It should be what you expected.

In SURFER's Grid routine we can ask for the square of this function,

$$[E(\xi, \eta)]^2 = (a^2)^2 \operatorname{sinc}^2(\pi a \xi) \operatorname{sinc}^2(\pi a \eta) = I(\xi, \eta) = \text{Irradiance}$$

Figure 7.5 is the irradiance pattern that would actually be seen as the Fraunhofer diffraction pattern of the square aperture. Notice that since a^2 is the area of the square aperture, the irradiance in the Fraunhofer diffraction pattern increases directly proportional to a^4, the square of the aperture area. The ξ- and η-axis scaling is unchanged. It was not intended for one to extract

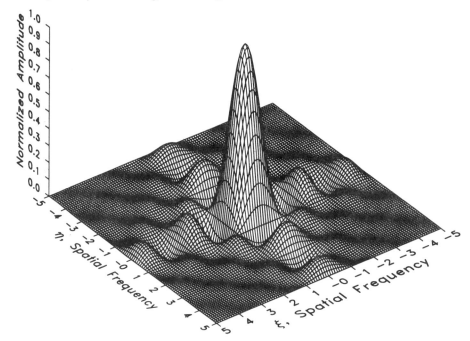

Figure 7.4 SURFER generated plot of $E(\xi, \eta) = \operatorname{sinc}(\pi a \xi) \cdot \operatorname{sinc}(\pi a \eta)$; normalized to 1 at $(0, 0)$.

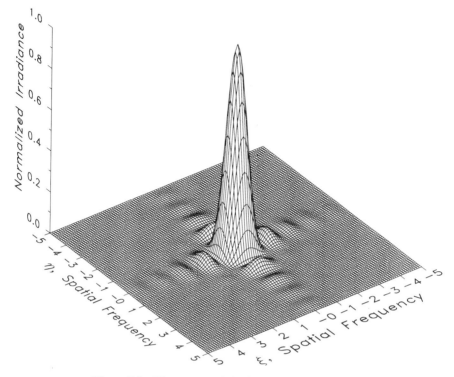

Figure 7.5 The square of the function of Figure 7.4.

quantitative data from these plots, but labeling the ξ- and η-axes does qualitatively provide information on spatial frequency content. We will often omit axis scaling.

If the reader is not content with the scales in terms of ξ and η, multiply each scale by:

- 2π to get u and v, the spatial angular frequencies;
- $R\lambda$ to get X or Y, actual distances in transform plane;
- λ to get $\sin\theta$ or $\sin\phi$, sines of the diffraction angles;
- πa to get α and β, single arguments of $E = (\text{sinc }\alpha)(\text{sinc }\beta)$ where: $\alpha = auX/2R$, $\beta = avY/2R$.

7.2.2 The Rectangular Aperture; Preparation for SURFER

If the aperture dimensions are not a-by-a, but a-by-b, rectangular, and if the y-dimension is b, then our starting integral would be

$$E(\xi, \eta) = \int_{-b/2}^{+b/2} \int_{-a/2}^{+a/2} A(x, y) e^{i2\pi(\xi x + \eta y)} \, dx \, dy$$

The amplitude pattern (Fourier transform) and irradiance pattern are easily seen to become:

$$E(\xi, \eta) = (ab)\,\text{sinc}(\pi a\xi)\,\text{sinc}(\pi b\eta)$$

$$[E(\xi, \eta)]^2 = I(\xi, \eta) = \text{Irradiance} = (ab)^2\,\text{sinc}^2(\pi a\xi)\,\text{sinc}^2(\pi b\eta)$$

Again we see the amplitude function is proportional to the aperture area, ab, while the irradiance is proportional to the square of the aperture area, $(ab)^2$. For the uniformly illuminated rectangular aperture here are the SURFER plots for the Fourier transform (Figure 7.6), and its square, the irradiance function (Figure 7.7).

The aperture width in x has been kept the same as for the square, a, but along the y-axis $b = 2a$. Since b is larger than a, the quantity $\text{sinc}(\pi b\eta)$ will hit its zeros, minima, and maxima sooner than those of the $\text{sinc}(\pi a\xi)$ function. Hence the pattern is spread more in the ξ-direction corresponding to the smaller aperture width a; see Figures 7.6 and 7.7. The function has shrunk along the η-axis corresponding inversely to the doubled value $b = 2a$. Shall we say there is more low spatial frequency content corresponding to the narrower width a? Certainly there is a different distribution in ξ than in η.

Without going through calculations it can be seen that rotation of the aperture in its plane will result in a one-to-one rotation of the diffraction pattern. So that if the aperture is rotated, say 30°, the diffraction pattern is unchanged except for an identical 30° rotation in its plane.

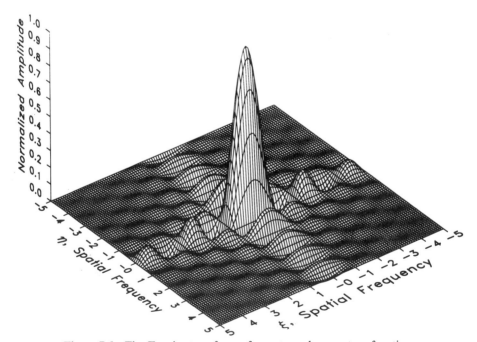

Figure 7.6 The Fourier transform of a rectangular aperture function.

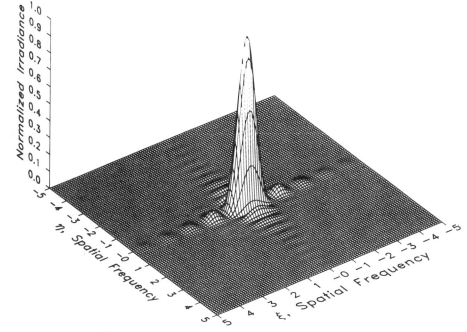

Figure 7.7 The square of the function of Figure 7.6.

7.2.3 Circular Aperture; Preparation for SURFER

The circular aperture, of radius a, is the first one we consider in which the mathematical derivation of the Fourier transform becomes a little more subtle. It is not our purpose to duplicate such derivations which are readily accessible in most optics textbooks. The calculation involves converting from Cartesian to spherical polar coordinates and then recognizing the presence of Bessel function relationships. Hecht explains this nicely (Hecht 1987). When this is done there results:

$$E(s) = (\text{constants})(2\pi a^2) \frac{J_1(2\pi a s)}{2\pi a s} \qquad (7.2)$$

The function, J_1, is the Bessel function of first order, the argument, $2\pi as$, uses the radial spatial frequency s which is analogous to the spatial frequencies ξ and η we have used up to now. They are related by

$$s^2 = \xi^2 + \eta^2$$

Just as $u = 2\pi\xi$ and $v = 2\pi\eta$, we define a radial *angular* spatial frequency, $\kappa = 2\pi s$. In so doing, the quotient for $E(s)$ would become that for $E(\kappa)$ and would have the form, without constant multipliers,

$$\frac{J_1(a\kappa)}{a\kappa}$$

In terms of an actual radial coordinate, Q, in the Fourier plane, we would write

$$\frac{J_1(2\pi a Q/\lambda R)}{(2\pi a Q/\lambda R)}$$

where $Q^2 = X^2 + Y^2$, X and Y are the Cartesian coordinates in the Fourier plane. If we plot E as a function of $\sin \theta$, we have

$$\frac{J_1[(2\pi a \sin \theta)/\lambda]}{[(2\pi a \sin \theta)/\lambda]}$$

The angle θ is that of diffraction, the angle in our Fraunhofer approximation whose sine is Q/R. We can let $2\pi as = \alpha$ and our E then becomes simply

$$J_1(\alpha)/\alpha$$

The first zero occurs when $\alpha = 3.83$, when $2\pi as = 3.83$; when $s = 3.83/2\pi a = 0.61/a = 1.22/D$; where $D = 2a$. But $s = Q/\lambda R$, hence the well-known relationship

$$Q = 1.22 \frac{\lambda R}{D} \quad \text{or} \quad 1.22 \frac{\lambda F}{D}$$

if we use the concepts explained in Chapter 5, letting R become F (the focal length of the lens following the circular aperture). We can go right to SURFER with $E(s) = J_1(2\pi as)/2\pi as$, setting $a = 1$, but type it in SURFER notation as

```
Z = (j1(6.283*sqrt(x*x + y*y)))/(6.283*sqrt(x*x + y*y))
```

since $s = \sqrt{x^2 + y^2}$. Again, a good choice for the range of s is 0 to 5, equivalent to ranging ξ and η through ± 5; amplitude and irradiance plots are given in Figures 7.8 and 7.9.

If one is plotting with software which does not make Bessel functions available, one can always resort to a polynomial approximation for them, though that can become somewhat unwieldy, carrying sufficient terms for a good approximation.

From the diffraction patterns for the square, rectangle, and circle, one may get the impression that no light is diffracted into the regions at larger angles away from the axes, regions of high spatial frequency. It appears that way on the E^2 plots, but the E plots indicate that there is light cast in those directions, weak though it may be. Gonzalez and Wintz (1987) have pointed out that the absolute value of E, $|E|$, is a useful plot. It shows the spectral amplitudes but phase information, $\pm E$ knowledge, is lost. Since the high spatial frequency information almost disappears in the E^2 plots, Gonzalez and Wintz suggest plotting $\log(1 + |E|)$. Though not the true irradiance function, it is similar and

SURFACE PLOTTING HAND-CALCULATED $Z(X, Y)$ 179

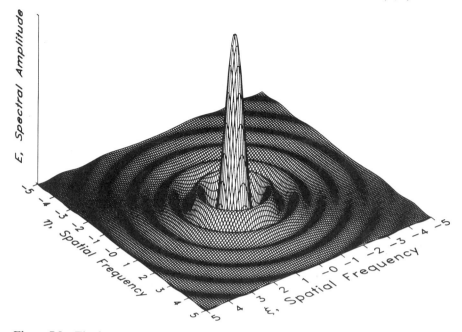

Figure 7.8 The form of $J_1(2\pi as)/(2\pi as)$, the Fourier transform of the circular aperture, $s^2 = \xi^2 + \eta^2$.

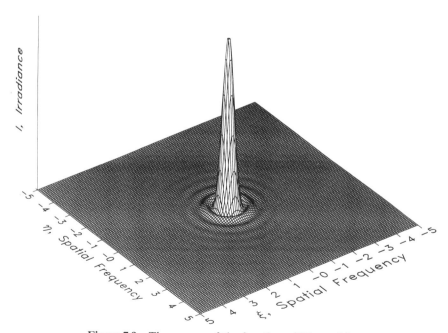

Figure 7.9 The square of the function of Figure 7.8.

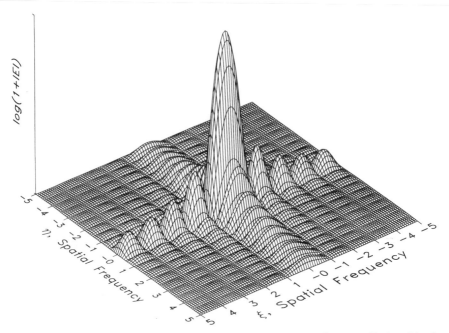

Figure 7.10 For the rectangular function's Fourier transform, $E(\xi, \eta)$, this is $\log(1 + |E(\xi, \eta)|)$.

allows one to see the high spatial frequency content. They indicate by example that this can be done photographically. It greatly enhances a photograph of a diffraction pattern without the over-exposure in the central, low frequency region.

To provide some indication of the amount of light at those higher spatial frequencies, here are some plots of $\log(1 + |E(\xi, \eta)|)$ for the rectangle and the circle (Figures 7.10 and 7.11). But one of the virtues of the plots of $[E(\xi, \eta)]^2$, the relative irradiance in 3-space, is that in the plots one can *see the real relative values of irradiance*; this is practically impossible to judge by eye or photograph; try it and see.

7.2.4 The Parallelogram Aperture; Preparation for SURFER

Let us now take the rectangular aperture and change it to the specific parallelogram aperture shown in Figure 7.12. We will work through some of the mathematics since we have not often seen this type of diffraction calculation in print. Though not a general derivation, the approach can be used with other apertures.

$$E(\xi, \eta) = \iint\limits_{\text{aperture}} A(x, y) e^{i2\pi(\xi x + \eta y)} \, dx \, dy$$

SURFACE PLOTTING HAND-CALCULATED $Z(X, Y)$ 181

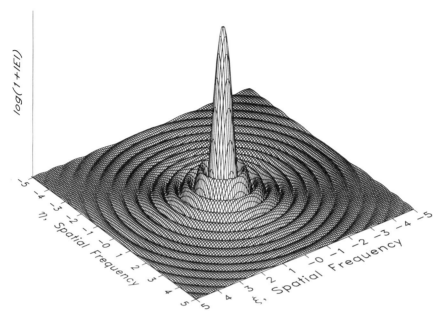

Figure 7.11 For the circular function's Fourier transform, $E(\xi, \eta)$, this is $\log(1 + |E(\xi, \eta)|)$.

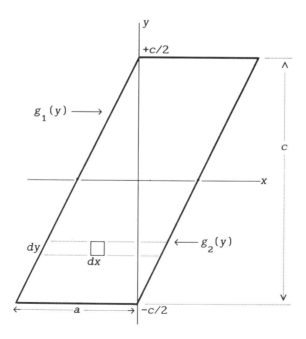

Figure 7.12 Diagram to represent an aperture of parallelogram shape.

This is the integral we must solve. The value of $A(x, y)$ is 1 inside the aperture and zero outside. The limits on the integrals will specify how they are to be evaluated, and those limits will constrain us to stay within the boundaries of the aperture. We'll integrate first over dx, from the left side which specifies x there as $g_1(y)$, to the right side where x is specified as $g_2(y)$. Then we do the second integration over dy from the line specified by $y = -c/2$ to the line where $y = +c/2$.

$$E(\xi, \eta) = \int_{y=-c/2}^{y=+c/2} \int_{x=g_1(y)}^{x=g_2(y)} (1) e^{i2\pi(\xi x + \eta y)} \, dx \, dy$$

Dimensionally, $c = 2a$. We will need the equations for the lines bounding the left and right sides at $x = g_1(y)$ and $x = g_2(y)$. In a parallelogram, the left and right sides have the same slope m, positive, and in this example $m = c/a$.

For $g_1(y)$:
$$y = mx + c/2 = (c/a)x + c/2 = y(\text{left})$$

For $g_2(y)$:
$$y = mx - c/2 = (c/a)x - c/2 = y(\text{right})$$

Solving the first for $x(\text{left})$ and the second for $x(\text{right})$:

$$x(\text{left}) = (a/c)y - a/2$$
$$x(\text{right}) = (a/c)y + a/2$$

Substituting into the integral,

$$E(\xi, \eta) = \int_{y=-c/2}^{y=+c/2} \int_{x=(a/c)y-a/2}^{x=(a/c)y+a/2} e^{i2\pi(\xi x + \eta y)} \, dx \, dy$$

$$E(\xi, \eta) = \int_{y=-c/2}^{y=+c/2} e^{i2\pi\eta y} \left[\int_{x=(a/c)y-a/2}^{x=(a/c)y+a/2} e^{i2\pi\xi x} \, dx \right] dy$$

$$E(\xi, \eta) = \int_{y=-c/2}^{y=+c/2} e^{i2\pi\eta y} \left[\frac{1}{i2\pi\xi} (e^{i2\pi\xi x}) \right]_{x=(a/c)y-a/2}^{x=(a/c)y+a/2} dy$$

Will we be able to do the integration in y?

$$E(\xi, \eta) = \int_{y=-c/2}^{y=+c/2} e^{i2\pi\eta y} \left(\frac{1}{i2\pi\xi} \right) \left(e^{i2\pi\xi(ay/c + a/2)} - e^{i2\pi\xi(ay/c - a/2)} \right) dy$$

$$E(\xi, \eta) = \int_{y=-c/2}^{y=+c/2} e^{i2\pi\eta y} \left(\frac{1}{i2\pi\xi} \right) \left(e^{i2\pi\xi ay/c} \cdot e^{i2\pi\xi a/2} - e^{i2\pi\xi ay/c} \cdot e^{-i2\pi\xi a/2} \right) dy$$

$$E(\xi, \eta) = \int_{y=-c/2}^{y=+c/2} e^{i2\pi\eta y} \left(\frac{1}{\pi\xi}\right)(e^{i2\pi\xi ay/c})\left(\frac{e^{i\pi\xi a} - e^{-i\pi\xi a}}{2i}\right) dy$$

$$E(\xi, \eta) = \left(\frac{1}{\pi\xi}\right) \sin(\pi a\xi) \int_{y=-c/2}^{y=+c/2} e^{i2\pi(\eta + \xi a/c)y} \, dy$$

Yes, the y-integral is an easy one.

$$E(\xi, \eta) = \left(\frac{1}{\pi\xi}\right) \sin(\pi a\xi) \left(\frac{1}{i2\pi(\eta + \xi a/c)}\right)(e^{i2\pi(\eta+\xi a/c)y}) \Big|_{y=-c/2}^{y=+c/2}$$

$$E(\xi, \eta) = \left(\frac{1}{\pi\xi}\right) \sin(\pi a\xi) \left(\frac{1}{\pi(\eta + \xi a/c)}\right) \left(\frac{e^{i2\pi(\eta+\xi a/c)c/2} - e^{-i2\pi(\eta+\xi a/c)c/2}}{2i}\right)$$

$$E(\xi, \eta) = \left(\frac{1}{\pi\xi}\right) \sin(\pi a\xi) \left(\frac{1}{\pi(\eta + \xi a/c)}\right) \sin[\pi c(\eta + \xi a/c)]$$

This is the Fourier transform for this specific parallelogram aperture **centered** on the x–y origin, in terms of its width, a, and vertical height, c. It is specific because the side slope is c/a, here specifically drawn and calculated with $c/a = 2$. If the quotients, $a/c = 1/$slope of the parallelogram, in these quotients only, if

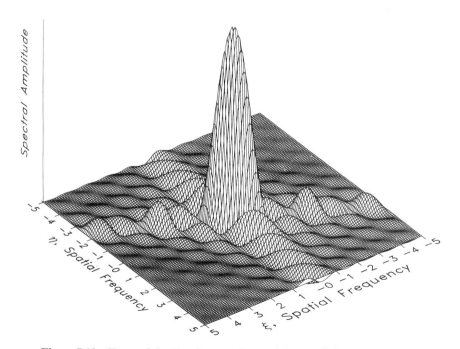

Figure 7.13 Form of the Fourier transform of the parallelogram aperture.

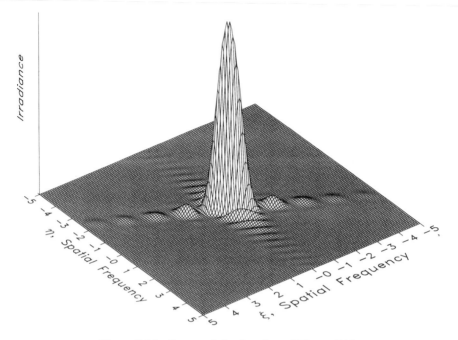

Figure 7.14 Square of the function of Figure 7.13.

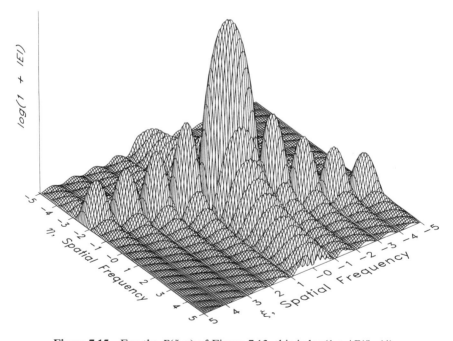

Figure 7.15 For the $E(\xi, \eta)$ of Figure 7.13, this is $\log(1 + |E(\xi, \eta)|)$.

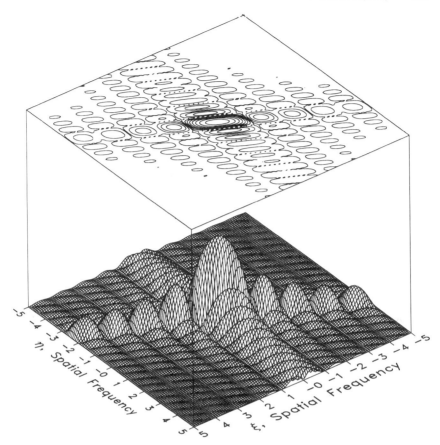

Figure 7.15(a) SURFER allows the stacking of several plots, and it can do contour maps.

$c = \infty$ it means vertical sides; a rectangle of sides a and c. In the quotients only if $c = \infty$, the expression becomes that for the Fourier transform of the rectangle. For the parallelogram, letting $a = 1$, $c = 2$, our expression becomes

$$E(\xi, \eta) = \left(\frac{2\sin(\pi\xi)}{\pi\xi}\right)\left(\frac{\sin(2\pi\eta + \eta\xi)}{2\pi\eta + \pi\xi}\right)$$

Now convert this so that it is suitable for SURFER.

$$z(x, y) = \frac{2\sin(\pi x)}{\pi x} \cdot \frac{\sin \pi(2y + x)}{\pi(2y + x)}$$

Since we are only concerned with relative values, the preparation for SURFER can omit the $2/\pi^2$ multiplier and we plot

$$z(x,y) = \frac{\sin(\pi x) \sin \pi(2y+x)}{(2xy + x \cdot x)}$$

That's not exactly SURFER format, but you get the idea. And SURFER only needs the part after the = sign. One needs to be careful that the correct math format is used in SURFER. It is easy to type in an improper expression. Figures 7.13, 7.14, and 7.15 are the corresponding plots for the parallelogram. In Figure 7.15(a), SURFER has allowed us to stack a contour plot of Figure 7.15 over itself.

7.3 LETTING THE SOFTWARE DO IT ALL; MATLAB DEVELOPMENT

Now let's examine a different approach using MATLAB, a matrix based math utility. MATLAB will accept our equation for the square's Fourier transform, the amplitude function, and plot it, but nothing new is to be gained by so doing. **Instead we'll use its matrix based math utilities** and *begin with the aperture function itself.*

7.3.1 The Square Aperture

As an example of the ease of use of MATLAB we'll first demonstrate diffraction again by a single square aperture. PC-MATLAB allows a 64 × 64 matrix working space; AT-MATLAB, which we have used here, allows a 90 × 90 matrix. Apparently, 386-MATLAB has no limit, but all have specific hardware requirements. Because of the reciprocal relationship between aperture size and spread of the Fourier transform, we find that a square of size 7 × 7 within the 90 × 90 matrix is a nice size for diffraction portrayal. We shall describe here each step for this square aperture but will not carry such detail throughout this book. Along the way we'll mention some of the unexpected "twists" of this software. So we'll let MATLAB begin with the actual square aperture, and let it also calculate the Fourier transform.

* * * * * *

```
sq = zeros(90);              creates a 90 × 90 array of zeros.
                             we chose the name "sq."
sq(43:49, 43:49) = ones(7);  creates the centered 7 × 7
                             square aperture; i.e.
                             sq(xstart:xend, ystart:yend).
mesh(sq)                     yields Figure 7.16.
sqft = fft2(sq);             creates the fast Fourier
                             transform.
mesh(sqft)                   will display it (not shown
                             here).
```

* * * * *

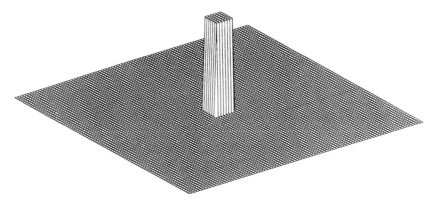

Figure 7.16 A MATLAB representation of a square aperture function with uniform amplitude illumination.

Very likely there will be a "spiky" nature to the meshed figure due to the nature of MATLAB's calculations, round-off's and some possible complexity due to phase elements. The first picture may not be very satisfactory and may look something like Figure 7.17. Also surprising is that the **fft2** algorithm places the zero order at one corner of the matrix, so each quadrant requires a 180 degree rotation.

Figure 7.17 MATLAB's **fft2** algorithm places the zero order in the matrix corners. The square aperture Fourier transform (real part) with zero order in the matrix corners.

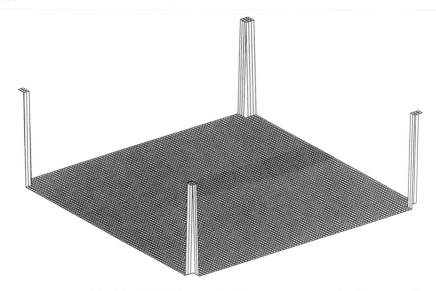

Figure 7.18 We'll fool MATLAB by putting the square aperture in the corners first, with **fftshift**.

To eliminate the spiky nature it seems best to *first shift the square*, rather than shifting the transform matrix, to the corners of the matrix, then do **fft2**. The command **fftshift**(sq); does that (Figure 7.18). Much of this can be combined into a single line, such as

```
sqreft=real(fftshift(fft2(fftshift(sq))));
mesh(sqreft)
```

yields Figure 7.19, the real part of the Fourier transform of the square aperture. It is sometimes referred to as the amplitude point spread function (ASF). Because some transforms are complex one might also wish to create the modulus. For this real function:

```
sqmdft=abs(fftshift(fft2(fftshift(sq))));
mesh(sqmdft)
```

yields the modulus (Figure 7.20).

With a 640 kbyte computer one often runs out of usable computer memory. With MATLAB the command **whos** will show what occupies memory. The command **pack** will rearrange it. Some few 100 kbytes are used in each of the calculations just performed in the 90 × 90 matrix. The command **clear** *filename*,

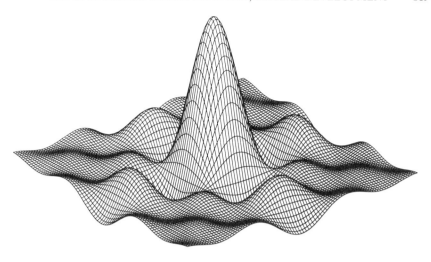

Figure 7.19 The square aperture Fourier transform (real part), zero order centered.

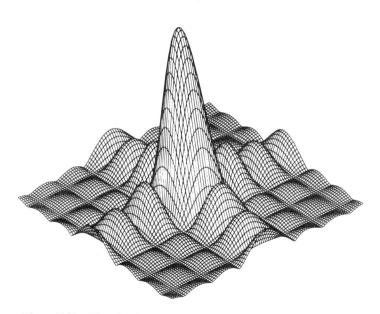

Figure 7.20 The absolute value of the function of Figure 7.19.

190 FOURIER OPTICAL TRANSFORMATIONS BY COMPUTER

followed by **pack** and **whos** will erase the named file, create more usable memory and show memory available. The command **save** *filename* will save the data to hard disk with the extension .mat, for reloading when needed again.

The Fraunhofer irradiance, also called the point spread function (*PSF*), can be displayed,

```
sqft=fftshift(fft2(fftshift(sq)));
```

this creates the amplitude function. It may be complex, and will be, if the aperture, sq, is not centered in the matrix. Recall that a shift from symmetry of a periodic function led to a phase factor in the function's Fourier series. A Fraunhofer diffraction pattern will be phase shifted also if the symmetric aperture is not centered on the optical axis; more will be said about this later. The command **mesh**(sqft) will display, and can only display the real part. The irradiance, then, is

```
psf=sqft.*conj(sqft);      .* is element-by-element
                           multiplication.
mesh(psf)                  will calculate and display
                           the point spread function
                           Figure 7.21).
```

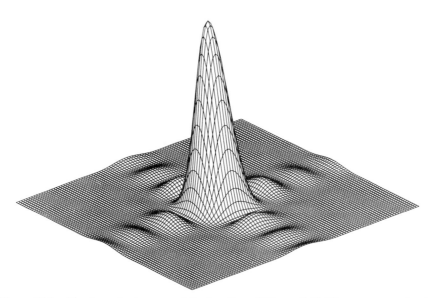

Figure 7.21 The (complex) square of the function of Figure 7.19. The square aperture's point spread function.

LETTING THE SOFTWARE DO IT ALL; MATLAB DEVELOPMENT

We know that the adjacent regions of light along the axes in the diffraction pattern of a square are consecutively phase shifted by π radians. Kaiser and Russell (1980) suggest a nice way to demonstrate this. But this phase shift does not occur diagonally in this pattern.

```
sqft=angle(sqft);      creates the phase matrix of the
                       aperture Fourier transform,
```

i.e.,

$$\arctan\left[\frac{\text{Imaginary part of sqft}}{\text{Real part of sqft}}\right]$$

The algorithm **angle** usually does not do a splendid job of keeping track of 2π phase changes. The command **unwrap** attempts to smooth out such phase plots by removing branch cuts in the arctan function. Several unwraps may bring continued improvement. We created a MATLAB function called **unwrap2** which executes the command **unwrap(unwrap(angle**(filename))$)'$' to speed up the operation, where apostrophes yield transposed matrices. Figure 7.22 shows phase in the Fourier transform of the square aperture, unwrapped a few times. The central maxima has zero phase, all steps are π (or multiples of π due to the unwrap function), and note the 2π, or zero, phase change diagonally.

Completion of the operations in Figure 7.2, bringing this application of the convolution theorem full circle, requires the calculation of the *optical transfer function (OTF)*. The *OTF* is very useful when the aperture function is used for imaging with incoherent light. The *OTF* is made up of a *modulation transfer*

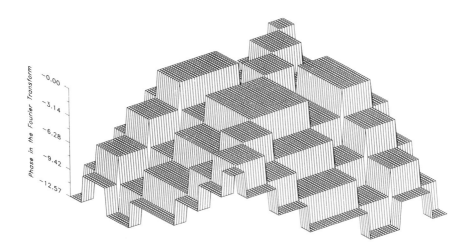

Figure 7.22 Phase in the Fourier transform of the square aperture.

192 FOURIER OPTICAL TRANSFORMATIONS BY COMPUTER

function (*MTF*) and a *phase transfer function* (*PTF*);

$$OTF = |MTF|e^{i(PTF)}$$

where $MTF = |OTF|$, the modulus of the *OTF*. The *MTF* describes the contrast allowed each spatial frequency in an image plane, while the *PTF* describes phase shifts for each spatial frequency in an image. This has been discussed in Chapter 6 and will be discussed in a later chapter.

The optical transfer function (*OTF*) can be calculated by two different paths according to Figure 7.2: (i) by inverse Fourier transformation of *PSF*, and (ii) by discrete convolution of the aperture function, which is *slower but more accurate*. If we do this with the *PSF* from the 7 × 7 aperture we will get a rather small *OTF* surrounded by a mostly flat surface of zeros. A 20 × 20 square can be created for method (i):

* * * * * *

```
sq=zeros(90);
sq(37:56,37:56)=ones(20);
sqft=fftshift(fft2(fftshift(sq)));
psf=sqft.*conj(sqft);
otf=ifft2(psf);
mesh(otf)
```
and we see it in the corners, quite spiky.

```
otf=fftshift(otf);
mesh(otf)
```
and small values of imaginary parts may have created phase noise. We know this *OTF* is real.

```
otf=abs(otf);
mesh(otf)
```
is a good portrayal (Figure 7.23).

* * * * * *

Having created the centered *PSF* as above, the *OTF* calculation can be condensed to a single line:

```
otf=fftshift(ifft2(fftshift(psf)));
```

this creates *OTF* without phase noise. Method (i) is limited to 45 × 45 squares in the 90 × 90 array of AT-MATLAB. Method (ii):

```
otf=conv2(ones(20),ones(20));
mesh(otf)
```
will portray a 30 × 39 gridded *OTF*.

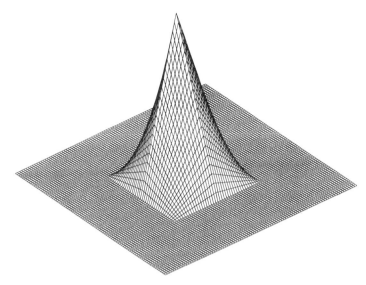

Figure 7.23 The optical transfer function of a square aperture obtained by inverse Fourier transformation of its point spread function.

The maximum size:

```
otf = conv2(ones(45), ones(45));
mesh(otf)
```
                                        yields an 89 × 89 gridded
                                        OTF, (Figure 7.24). It may
                                        take minutes.

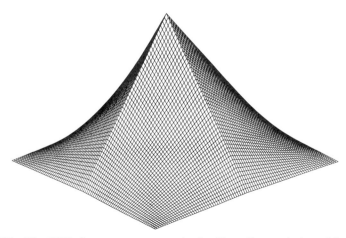

Figure 7.24 The *OTF* of a square aperture, obtained by self-convolution of the aperture function.

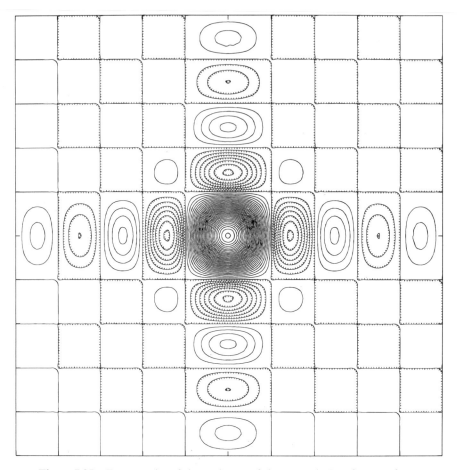

Figure 7.25 Topography of the real part of the square's Fourier transform.

MATLAB can output to a printer and figure tilt and rotation can be changed within MATLAB. We often translated MATLAB data files to SURFER binary Grid files and plotted out of SURFER. SURFER as well as MATLAB can do topographic plots. Figure 7.25 is the topography of Figure 7.4.

7.3.2 Other Real Apertures

We are impressed by the simplicity now available for 3-space calculation and display to simulate Fourier diffraction techniques. We will now demonstrate a few more. By *real* apertures we mean those which contain no phase altering structure.

7.3.2.1 Half Square, Half Transmitting An interesting variation on the square aperture is the square aperture with half of it only partially transmitting.

LETTING THE SOFTWARE DO IT ALL; MATLAB DEVELOPMENT 195

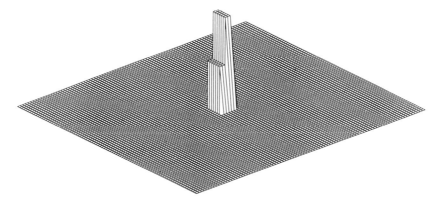

Figure 7.26 Square aperture, one half with only $\frac{1}{2}$ amplitude transmittance.

Using the same 7 × 7 square in MATLAB and reducing amplitude transmission by $\frac{1}{2}$ in one half, the resulting aperture is shown in Figure 7.26, which is a good size for Fourier transforming. Figure 7.27 is a better illustration; the transform can be calculated by hand and all parts put into SURFER. The modulus of the Fourier transform (Figure 7.28) looks distinctly different. Phase changes are shown in Figure 7.29 and the topography of the modulus is seen in Figure 7.30. Figure 7.30(a) is the real part of the transform; Figure 7.30(b) is the imaginary part; Figure 7.30(c) is the irradiance function; Figure 7.30(d) is the convolution of Figure 7.27, the aperture with itself.

Comparison with the fully transmitting square is worthwhile, and further transmission reduction of half of the aperture, say to $\frac{1}{4}$, leads one to what would be expected, thinking of reduction all the way to zero. In this case, zero, the aperture would no longer be symmetric and a linear phase shift will appear

Figure 7.27 A better size portrayal of the aperture of Figure 7.26, done in MATLAB; it is too large to give a good Fourier transform.

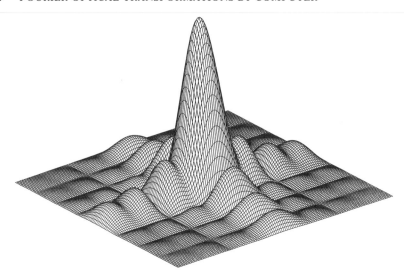

Figure 7.28 The aperture of Figure 7.27 is not symmetric; its Fourier transform, $E(\xi, \eta)$, will be complex valued at each point (ξ, η). This is the absolute value of $E(\xi, \eta)$ or the modulus of $E(\xi, \eta)$.

in one direction. These formulas can be calculated by hand, and SURFER used to get higher resolution than 90 × 90 plots; 126 × 126 gives nice detail. Care to try the aperture with half only $\frac{1}{4}$ transmitting?

7.3.2.2 Circular Diffracting by a circular aperture can be demonstrated by the same procedures. It is impossible to get a perfectly round circle in a 90 × 90

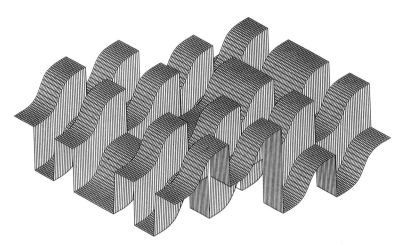

Figure 7.29 Phase in the Fourier transform of the function of Figure 7.26.

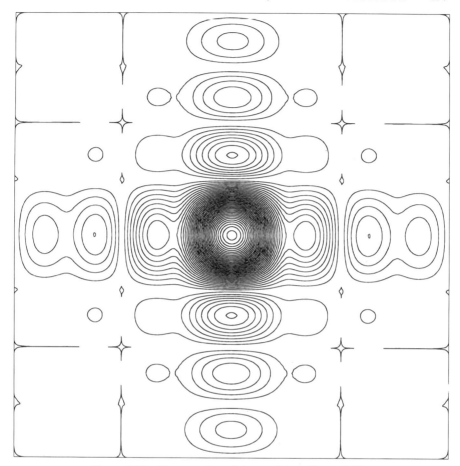

Figure 7.30 Topography of the modulus, Figure 7.28.

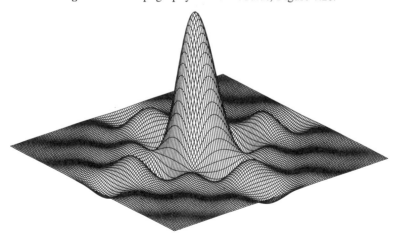

Figure 7.30(a) The real part of the Fourier transform of the function of Figure 7.26.

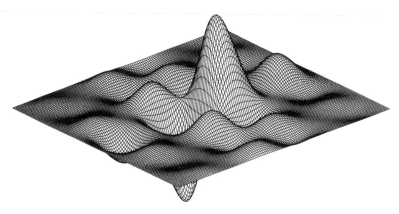

Figure 7.30(b) The imaginary part of the Fourier transform of the function of Figure 7.26.

square matrix of MATLAB, especially of small enough size with which to do good Fourier transforms; the ragged edge affects the transforms noticeably. We created a MATLAB function called **circ**(radius) to produce square-symmetric circular apertures. SURFER understands Bessel functions of the first and second kind, of any order, and will yield high resolution graphic simulations when working with circular apertures. SURFER's menus are easy to use, **but SURFER cannot do Fourier transforms, convolutions or correlations**.

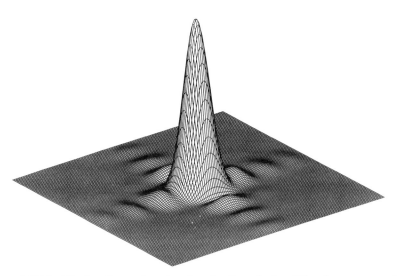

Figure 7.30(c) The irradiance function for the Fraunhofer diffraction pattern of the aperture of Figure 7.26. The (complex) square of the Fourier transform of the function of Figure 7.26.

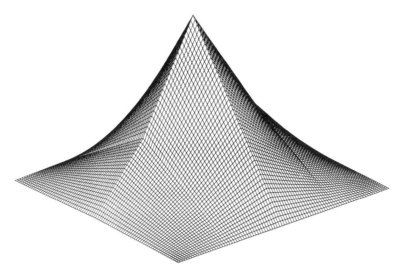

Figure 7.30(d) The self-convolution of the aperture of Figure 7.27.

7.3.2.3 Square and Circle, a Composite Aperture Created in MATLAB back in Chapter 6, Figure 6.21(a) is the double aperture, the convolution and auto-correlation of which was discussed in Section 6.4; i.e., a square and circle each of the same width, 24, spaced similarly. This is how the aperture was created in MATLAB

* * * * * *

```
ap=circ(12);
ap(49:72,:)=ones(24);
db=zeros(74,26);
db(2:73,2:25)=ap;
mesh(db)                    Aperture display with faulty
                            screen perspective?
otf=conv2(db,db);
mesh(otf)                   displays the OTF. When plotted
                            the perspective was correct.
```

* * * * * *

The following command did the auto-correlation:

```
sqcrcorr=xcorr2(db,db);     and then, mesh(sqcrcorr)
```

7.3.2.4 Triangle

An equilateral triangle of appropriate size for transformation can be created:

* * * * * *

```
ap = zeros(89);
for i = 35:45
jmin = (6/10)*(i-45) + 45;
jmax = -(6/10)*(i-45) + 45;
for j = jmin: jmax
ap(i,j) = 1;
end;
end;
mesh(ap)                                  yields Figure 7.31
                                          and the matrix for
                                          transformations.
ft = fftshift(fft2(fftshift(ap)));        for the Fourier
                                          transform.
```

* * * * * *

MATLAB returns the real part of the transform on **mesh**(ft); or use reft = **real**(ft); and **mesh**(reft). The command **abs**(ft) returns the modulus; **imag**(ft) returns the imaginary part. Figure 7.32 is the real part. The command **angle**(ft) and a few unwraps will return a phase plot of the transform, such as Figure 7.33. After translating the reft matrix to a grid file for SURFER, the topography of the real part was plotted (Figure 7.34).

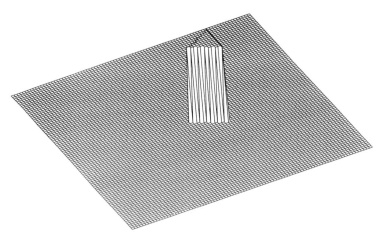

Figure 7.31 A MATLAB representation of a triangular aperture.

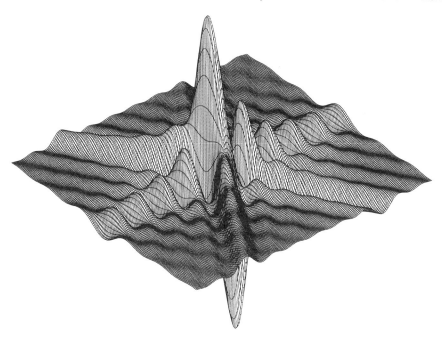

Figure 7.32 The real part of the Fourier transform of the function of Figure 7.31.

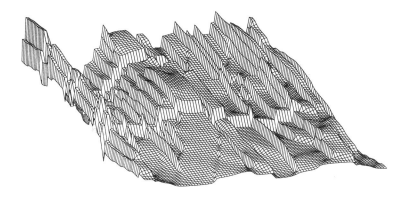

Figure 7.33 Phase in the Fourier transform of the function of Figure 7.31; all steps are π.

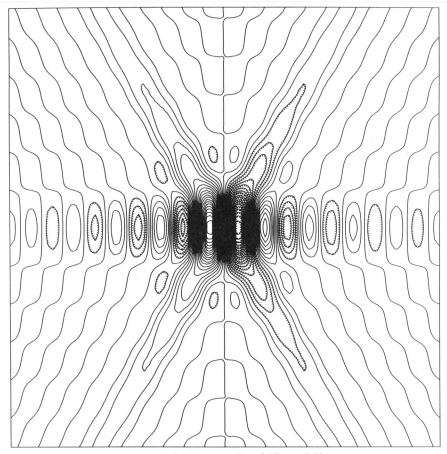

Figure 7.34 Topography of Figure 7.32.

For the *OTF* a larger triangle can be established:

* * * * * *

```
ap=zeros(40);
for i=1:39
jmin=(1/sqrt(3))*(i-39)+23;
jmax=-(1/sqrt(3))*(i-39)+23;
for j=jmin:jmax
ap(i,j)=1;
end;
end;
mesh(ap)                              to see it as Figure 7.35.
triotf=conv2(ap, ap);
mesh(triotf)                          yields Figure 7.36, the
                                      OTF.
```

* * * * * *

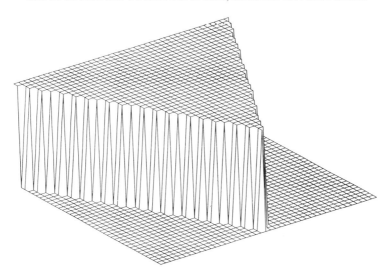

Figure 7.35 A larger triangular aperture, in preparation for its self-convolution to create the *OTF*.

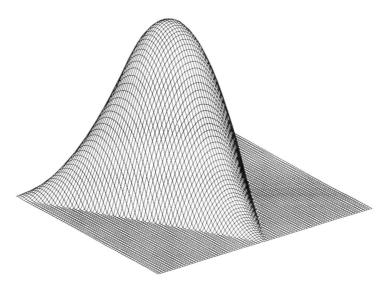

Figure 7.36 The self-convolution of the function of Figure 7.35.

This last calculation seems slow even on a 386 computer. Smith and Marsh (1974) have shown how to do some of these calculations by hand for the triangular aperture. Figures 7.37 and 7.38 show the modulus of the transform, and the irradiance function.

7.3.2.5 Composite Apertures, Annular Diffraction by composite apertures can easily be simulated because of the linearity of Fourier transforms. The

204 FOURIER OPTICAL TRANSFORMATIONS BY COMPUTER

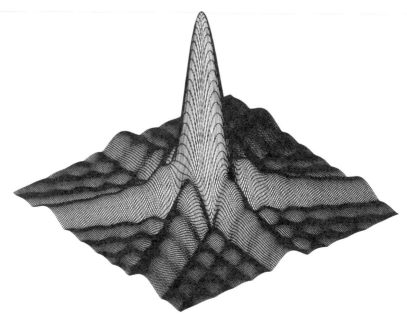

Figure 7.37 For the triangular aperture of Figure 7.31, this is the modulus of its Fourier transform.

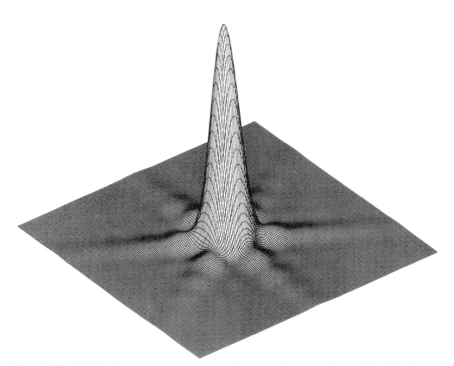

Figure 7.38 This is the (complex) square of the Fourier transform of the function of Figure 7.31, the Fraunhofer irradiance.

LETTING THE SOFTWARE DO IT ALL; MATLAB DEVELOPMENT 205

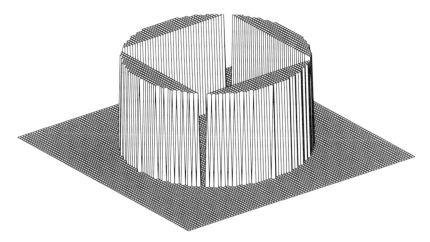

Figure 7.39 Circular aperture blocked by an opaque square.

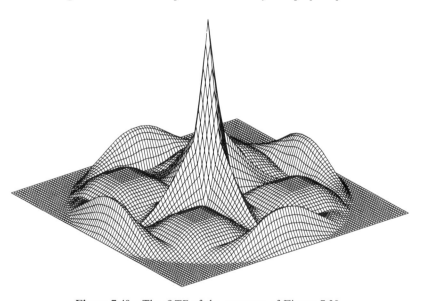

Figure 7.40 The *OTF* of the aperture of Figure 7.39.

diffraction amplitude of an annular ring can be determined by subtracting the transform of the interior circle from that of the exterior circle; diffraction by an annular square follows similarly. Since these calculations can be done by hand, SURFER can yield high resolution graphics from them. But using the 90 × 90 array of MATLAB, Figure 7.39 shows a circular aperture blocked by an opaque square of diagonal the same as the circle's diameter. Figure 7.40 shows the real optical transfer function obtained by discrete convolution of the aperture, a calculation one would not like to do by hand. In Figure 7.41,

Figure 7.41 Topography of Figure 7.40.

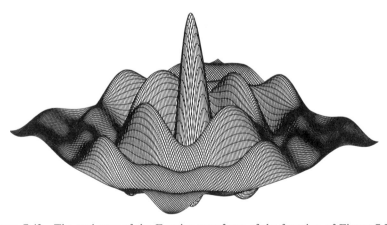

Figure 7.42 The real part of the Fourier transform of the function of Figure 7.39.

SURFER has done the topography of Figure 7.40. Figures 7.42, 7.43, and 7.44 show the transform's real part, modulus, and complex square. Beautiful, yes?

Note that earlier, for a one-dimensional function $f(x)$ we were able to use for $f(x)$ both negative and positive values (bipolar). Now, the real part of our present two-dimensional aperture functions, $A(x, y)$, can only be positive; negative transmission has no physical meaning. *Real apertures cannot have*

Figure 7.43 The modulus of the function of Figure 7.42.

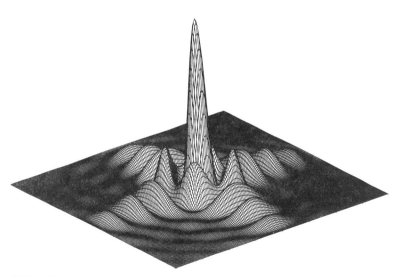

Figure 7.44 The (complex) square of the Fourier transform of the function of Figure 7.39, the Fraunhofer irradiance.

transmission of odd symmetry. Mathematically, however, two-dimensional real functions of odd symmetry can be allowed.

FOR FURTHER CONSIDERATION

1. In the positive range, $x =$ zero to 5, the $\sin(\pi x)$ function has three maxima at $x = \frac{1}{2}, \frac{5}{2}, \frac{9}{2}$. The $\text{sinc}(\pi x)$ function, nothing more than $\sin(\pi x) \cdot (1/\pi x)$, the $1/\pi x$ being a factor to diminish the amplitude, has only two maxima for $x =$ zero to 5, and one at the origin. What happened to the $\sin(\pi x)$ maximum at $x = \frac{1}{2}$?

2. This too, for consideration: the interested reader might possibly find useful the book *Fast Fourier Transforms* by James S. Walker, CRC Press, 1991, with its Chapter 6, "Fourier Optics," and its computer disk for doing fast Fourier transforms in one-dimension.

    There is also a generally useful, somewhat inexpensive, software package for technical plotting and data processing called PSI-PLOT, which can do Fourier transforms of one-dimensional data. It is sold by Poly Software International, P.O. Box 526368, Salt Lake City, UT 84152, USA.

SURFER: Golden Software, 809 14th Street, Golden, CO 80401-1866, USA.

MATLAB: The MathWorks, Inc., 24 Prime Park Way, Natick, MA 01760-1500, USA.

Another software package apparently similar to MATLAB, capable of Fourier transforms and convolutions on two-dimensional data arrays is O-MATRIX available from:

Harmonic Software, Inc., 12223 Dayton Avenue, North Seattle, WA 98133, USA.

As with all computer software, you are advised to try before you buy.

8

APODIZATION AND SUPER-RESOLUTION, PHASE FROM SHIFT, AND MULTIPLE APERTURES

The diffraction process, as we have been considering it, is the *spreading* of light as it passes through apertures or past obstructions. In optical instruments such as telescopes and microscopes this *spreading* is generally not desirable. The best example which comes to mind: starlight which is to be processed by a telescope should arrive at the telescope aperture as a plane wave. (At the aperture of even our largest optical telescopes, due to the star's distance, many light years, there will be an undetectable geometrical wave-front curvature due to distance.) Incident plane waves; this is the condition for Fraunhofer diffraction. Of course, atmospheric turbulence and refraction will continuously make the wave front other than plane. (But see Chapter 11.) Because of their great distance, essentially infinite, such stars cannot display in the telescope any perceptible angular size; they should appear as a point of light in the focal plane of the telescope. The focal plane becomes the Fourier transform plane. The star image is proportional to the square of the Fourier transform of the telescope aperture (or of the wave front at the exit pupil, see Chapter 6), displayed compositely in all visible wavelengths being received from the star. (The giant star Mira, Omicron Ceti, at a distance of 228 light years, though 420 times the size of the sun has an angular diameter, the angular size as viewed from earth, of only 0.056 seconds of arc; this size has been determined by other than direct means. It (the "disk" of Mira) cannot be resolved even by the 200-inch Palomar telescope but part of the problem is atmospheric turbulence.)

Instead of a nice crisp point of light, on film say, one gets a central maximum irradiance "surrounded by" various secondary maxima and minima, the pattern of which has nothing to do with the star and, unless one looks very carefully, is essentially the same for all stars of similar brightness. The shape of the image is

210 APODIZATION AND SUPER-RESOLUTION

characteristic of the aperture (or exit pupil wave front) of the telescope, not the star. Looking at photographs of stars or star fields, the images have been formed by diffraction in the telescope and one can often see in the brightest images diffraction evidence of some structures in the aperture, e.g., a secondary mirror, it's supports ("spiders"). etc. The diffraction effects are very evident in the photograph of the star shown in Figure 8.1(a). Christmas card artists may have been the first to take liberties with such photographs, but why a NASA painting would include diffraction effects remains a mystery to us (Figure 8.1(b)).

Your own visual instruments (eyes) have apertures which have a somewhat "ragged" edge, the inner edge of the iris. Distant points of light viewed at night do not appear to have smooth or sharp boundaries on their central maxima, but instead the light seems to flare, streak or smear out. Much of this is caused just by the "ragged" edge of the iris aperture. Put a strong magnifying glass in contact with a mirror and look in to examine your "ragged" edge iris; it needs some illumination from the side. It is not as raggedy as one might think.

It seems then that an imaging system will never be perfect because diffraction effects will always be present; excellent optics are said to be (only) "diffraction limited." If diffraction effects were not present, then the image of a star would be a tiny single point of light and the irradiance as we would plot it would be a single narrow spike on the optical axis. But instead, for all imaging devices, the central region of the Fraunhofer diffraction pattern in general will be spread

Figure 8.1(a) From the 200-inch Hale telescope of Mt. Palomar, a star image showing diffraction from the telescope secondary mirror supports and other aperture structures. (From A. Sandage, *The Hubble Atlas of Galaxies*, Carnegie Institution of Washington, Publication 618, an incidental star from Atlas page 24, 1961. Courtesy of Carnegie Institution of Washington; reprinted by permission of Palomar/California Institute of Technology.)

Figure 8.1(b) Diffraction spikes in a painting? What was the artist looking through? (From R. O. Fimmel, W. Swindell, and E. Burgess, *Pioneer Odyssey*, National Aeronautics and Space Administration, NASA SP-396, p. 138, 1977.)

out and there will be various secondary maxima and minima diffracted away from the central region. We've seen this in all our examples so far. This raises the interesting question, "Can we design apertures to minimize diffraction effects?" Or, "If we can't get perfect images, can we get 'best' images?" But then, what constitutes a best image? Can we design an aperture to diffract very little light and can we also narrow down the central maximum?

8.1 APODIZATION AND SUPER-RESOLUTION

Apodol, adj: [⟨Greek: apous, apodos, footless] Zool. - lacking feet or legs such as a snake or legless lizard.

Apodization, n. {by use of -ization, a n.-forming compound suffix} - the act of removing feet and/or legs. "I'd like a drumstick. Please apodize the turkey."

Can we narrow the central maximum and diffract very little light? The answer to both is yes, but the answer seems also to be: not both at the same time. The term apodization has been rather loosely used by some authors to mean improved resolution; improved resolution would suggest a narrowing of the central maximum, trying to approach a "spike." Some writers suggest that the annular ring aperture, which certainly narrows the central maximum, is an

212 APODIZATION AND SUPER-RESOLUTION

example of apodization. But clearly the secondary maxima, the feet, have not been suppressed—have not been apodized. Indeed, this approach enhances the secondary diffracted light. Figure 8.2 is the diffracted irradiance from a circular operture, while Figure 8.3 is the irradiance function from a circular ring aperture. The term *apodization* should be reserved to mean the diminishing, or removing, of the secondary diffracted light.

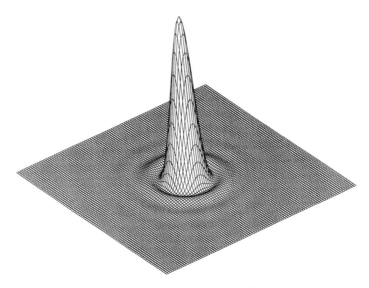

Figure 8.2 The Fraunhofer diffracted irradiance from a circular aperture.

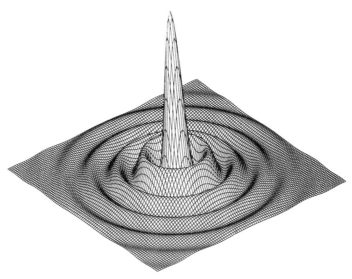

Figure 8.3 The Fraunhofer diffracted irradiance from a circular ring aperture.

APODIZATION AND SUPER-RESOLUTION   213

Figures 8.2 and 8.3 were done in SURFER by very careful use of equation (7.2) in which you can note that the aperture radius shows up as a constant factor. Using linearity, we subtracted the transform of a 0.4 radius circle function from the transform of a 0.5 radius circle function to get $E(\xi, \eta)$, the transform for the ring aperture function. The square of this $E(\xi, \eta)$ is Figure 8.3, while Figure 8.2 is the irradiance for the open 0.5 radius circular aperture. Figure 8.3(a)

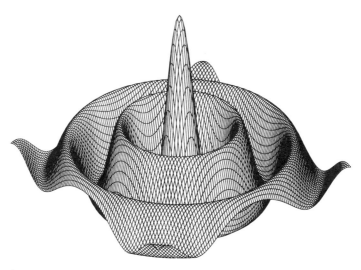

Figure 8.3(a) The Fraunhofer amplitude of the circular ring aperture.

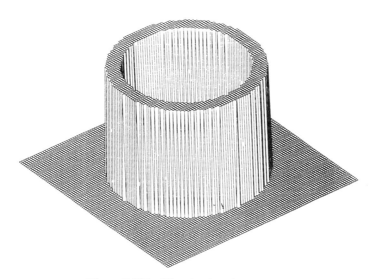

Figure 8.3(b) The circular ring aperture.

214 APODIZATION AND SUPER-RESOLUTION

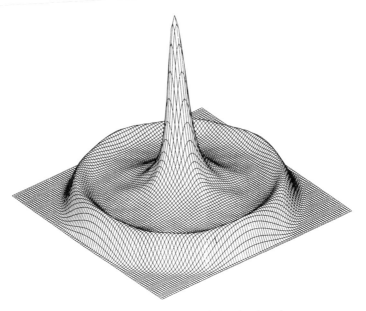

Figure 8.3(c) The self-convolution of the circular ring aperture.

is $E(\xi, \eta)$ for the ring aperture; sort of wild and woolly, yes? Figure 8.3(b) is the ring aperture function and Figure 8.3(c) is its convolution, the *OTF*.

Narrowing of the central maximum is usually a basis of the resolution criterion referred to as Rayleigh, Sparrow, or Dawes. When one tinkers with the aperture function to decrease the width of the diffraction central maximum smaller than that of the clear aperture, we refer to this as *super-resolution*, keeping faith with Rayleigh, who used the width of the central maximum as a criterion.

We have found that in general, *for mathematically real apertures*, if the aperture transmission function decreases from center to edge, such a function acts as an apodizer; the secondary maxima of irradiance in the diffraction plane will be suppressed. The suppression will be accompanied by a broadening of the central maximum. In the inverse situation, with the aperture transmission function decreasing from edge to center, the Fraunhofer irradiance is inversely affected; the secondary maxima are increased but the central maximum is narrowed. The annular ring aperture does this; so does a double slit. In either case, using such transmission functions opens the possibility of higher resolution in certain specific imaging situations.

8.1.1 Apodization

8.1.1.1 Cosine Apodization   It is easy to demonstrate the effect of apodization by using a square aperture and altering the amplitude transmission function in

APODIZATION AND SUPER-RESOLUTION 215

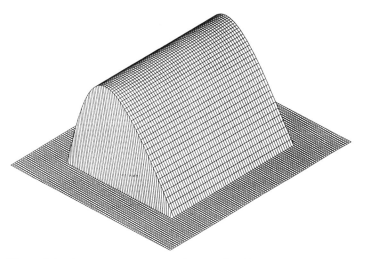

Figure 8.4 A square aperture, cosine apodized in the y-direction only.

only one direction. We'll apodize the aperture with a cosine function, taper the transmission from the center to the edge in the y-direction with $A = \cos(\pi y/a)$. The transmission goes to zero at the y edges of the aperture where $y = \pm a/2$. Figure 8.4 shows this aperture function. The Fraunhofer amplitude is the Fourier transform of that function:

$$E(\xi, \eta) = \int_{-a/2}^{+a/2} \int_{-a/2}^{+a/2} \cos(\pi y/a) e^{-i2\pi(\xi x + \eta y)} \, dy \, dx$$

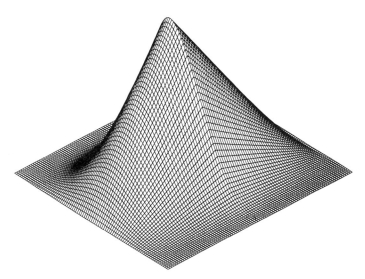

Figure 8.5 The self-convolution of the aperture function of Figure 8.4.

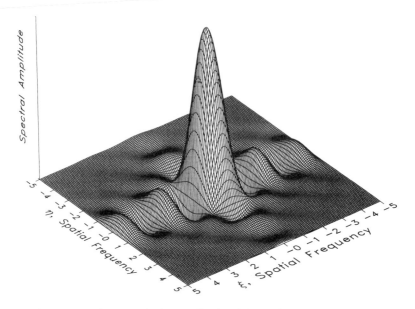

Figure 8.6 The Fraunhofer amplitude of the cosine apodized aperture.

The solution of this integral, with $a = 1$, is

$$E(\xi, \eta) = \frac{\sin(\pi\xi) \cos(\pi\eta)}{(\pi^2 \xi)(1 - 4\eta^2)}$$

Figure 8.5 shows the convolution of this real aperture function with itself; it is the optical transfer function (*OTF*) for imaging with this aperture and incoherent light. The Fraunhofer amplitude function is shown in Figure 8.6, and its square (Figure 8.7) is the irradiance or point spread function (*PSF*) for this aperture. Wow! How's that for removing secondaries? In the *PSF* note that though the secondary maxima are unchanged in the ξ-direction, we have greatly reduced them in the η-direction. This is what apodization is all about. But note from the topography of the $E(\xi, \eta)$ function (Figure 8.8) where the straight lines are at the zero level, that the central maximum for this square aperture is now larger in the η-direction than in the ξ-direction. We achieved apodization in η but broadened the central maximum in that direction.

8.1.1.2 Linear Taper Apodization A rather interesting example of an apodizing aperture is one with a linear decrease of amplitude transmission from center to edge. Figure 8.9 shows this decrease for a square aperture, with the decrease only in the *x*-direction so that we can compare diffraction in

APODIZATION AND SUPER-RESOLUTION   217

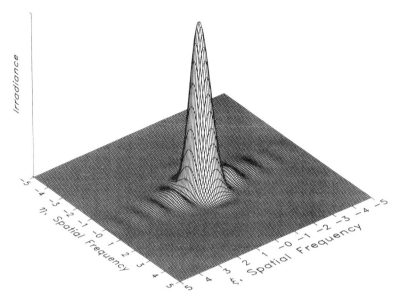

Figure 8.7 The Fraunhofer irradiance of the cosine apodized aperture. We chose to apodize only in one direction so that comparison could be made with the unaltered direction.

both ξ- and η-directions. Solving for $E(\xi, \eta)$ requires the solution of this Fourier transform,

$$E(\xi, \eta) = \int_{-a/2}^{+a/2} e^{-i2\pi\eta y}\, dy \left[\int_{-a/2}^{0} \left(\frac{2}{a}x + 1\right) e^{-i2\pi\xi x}\, dx \right.$$
$$\left. + \int_{0}^{+a/2} \left(\frac{-2}{a}x + 1\right) e^{-i2\pi\xi x}\, dx \right]$$

The y part separated out. The expressions in parentheses are the up and down sloping amplitude transmissions on either side of x equal zero; transmission is equal to one at $x = 0$. We must admit that solution using the exponential transform became a bit messy. But toward the end, if you keep in mind

$$e^{\pm i\theta} = \cos\theta \pm i\sin\theta$$

and

$$2\sin^2\theta = 1 - \cos 2\theta,$$

solution can follow.

$$E(\xi, \eta) = a\,\text{sinc}(a\pi\eta) \cdot \frac{a}{2}\,\text{sinc}^2\!\left(\frac{a\pi\xi}{2}\right)$$

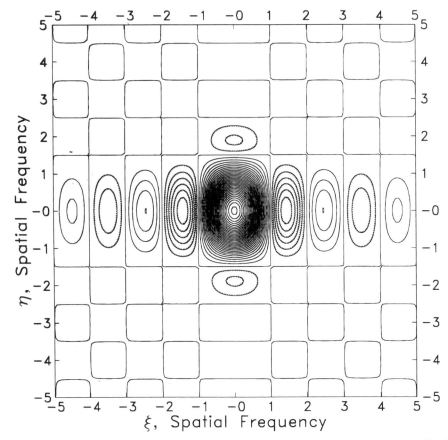

Figure 8.8 Topography of the Fraunhofer amplitude function for the cosine apodized aperture, apodization apparent in the η-direction.

What is somewhat surprising about the result is that this linear decrease of aperture amplitude transmission in x caused the amplitude function in ξ, the Fourier transform function, to be squared and therefore always positive; that's something you don't expect to see until the irradiance function.

Figures 8.10 and 8.11 show the amplitude and irradiance functions; in the irradiance function, comparing along ξ- and η-directions, you can see effective apodizing in ξ but also an increased width of the central maximum in ξ.

How did Mother Nature manage to get all the $E\,dS$ contributions from the aperture to always add positively here in ξ? A linear decrease in the y-direction would also give an aperture amplitude transmission function that looked like a tent. How could one create such a transmission aperture, for real, in the lab? We wonder also.

Solution for $E(\xi, \eta)$ might be easier using the proper trigonometric Fourier transform for this even aperture function.

APODIZATION AND SUPER-RESOLUTION 219

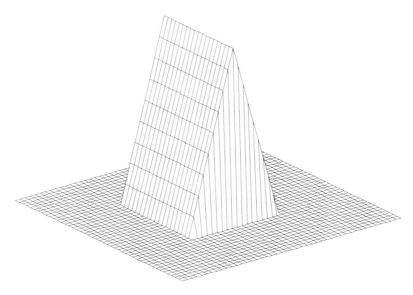

Figure 8.9 Aperture with a linear taper amplitude transmission function in the x-direction, the apodization direction.

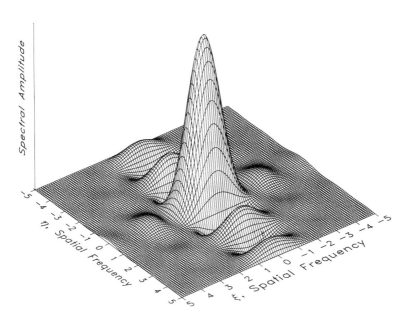

Figure 8.10 Fraunhofer amplitude for the linear taper apodization, effect apparent in the ξ-direction.

220 APODIZATION AND SUPER-RESOLUTION

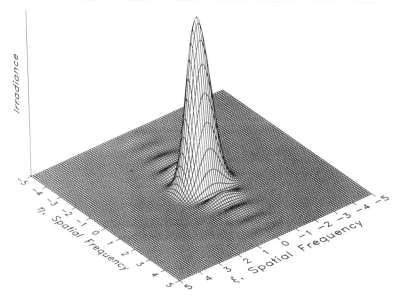

Figure 8.11 Fraunhofer irradiance for the linear taper apodization, effect apparent in the ξ-direction.

8.1.2 Super-Resolution

8.1.2.1 Cosine Super-Resolution The inverse of the apodization operation should provide a super-resolving aperture. This time we'll alter the aperture function with a negative cosine function. Recall that there is no meaning to negative transmission. Our aperture will taper in transmission from maximum at the y edges to zero in the center, only along the y-axis. The transmission function is $A(x, y) = 1 - \cos(\pi y/a)$ (Figure 8.12). The Fraunhofer amplitude is this Fourier transform

$$E(\xi, \eta) = \int_{-a/2}^{+a/2} \int_{-a/2}^{+a/2} (1 - \cos(\pi y/a)) e^{-i2\pi(\xi x + \eta y)} \, dy \, dx$$

It has a solution (Figure 8.13), again with $a = 1$,

$$E(\xi, \eta) = \text{sinc}(\pi\xi) \left\{ \text{sinc}(\pi\eta) - \frac{\cos(\pi\eta)}{1 - 4\pi\eta^2} \right\}$$

The topography of $E(\xi, \eta)$ (Figure 8.14), straight lines at the zero level, shows a small but measurable narrowing of the central maximum in the η-direction—a Rayleigh criterion to better resolution. But the increased irradiance in the secondaries (Figure 8.15) is obvious. Damn! The *OTF* for this super-resolving

APODIZATION AND SUPER-RESOLUTION   221

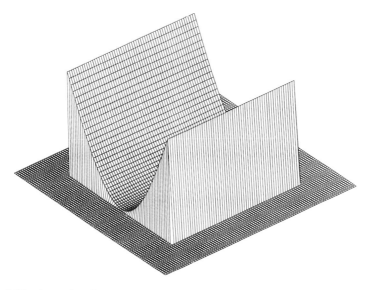

Figure 8.12 A cosine function used to create a super-resolving aperture. It will narrow the central maximum but watch for increased irradiance in the secondaries.

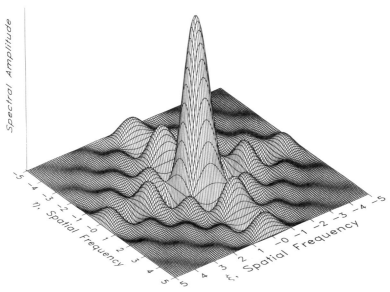

Figure 8.13 The Fraunhofer amplitude for the cosine super-resolver, effect not too apparent here.

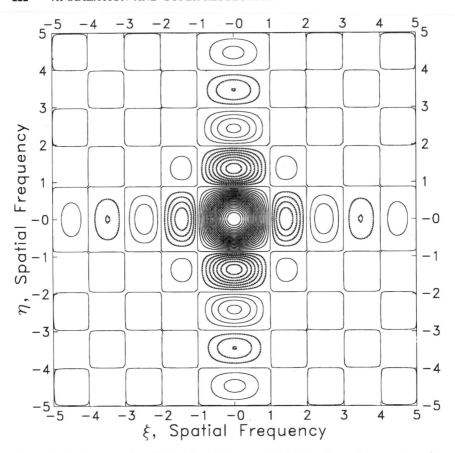

Figure 8.14 Topography of the Fraunhofer amplitude for the cosine super-resolver reveals the central maximum narrowed in the η-direction.

aperture is shown in Figure 8.16. Do you understand why it looks somewhat, but not exactly, like a double-slit convolution?

Consider two distant point objects very close together, the aperture and lens attempting to image the two points as separate. The two diffraction patterns overlap. If one image centers on a secondary maximum of the other there will be difficulty resolving the two. So for better resolution it might be helpful to *apodize* the aperture to suppress the secondary maxima and get them out of the way, but in so doing (apodizing) both central maxima expand and *they* may now overlap, also giving poor resolution. The one aperture is imaging both objects.

Trying super-resolution, shrinking the pattern toward the center, it is possible the increased irradiance in the secondaries might damage resolution. Perhaps we need "tailor-made" apertures, or better yet, dynamically variable apertures. We will describe "adaptive optics" in Chapter 11. For the two close stars

APODIZATION AND SUPER-RESOLUTION 223

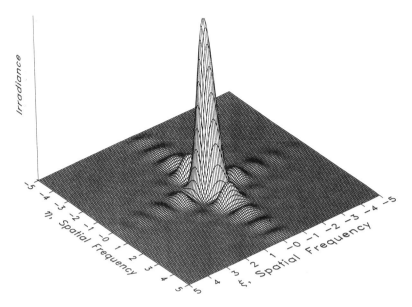

Figure 8.15 The Fraunhofer irradiance for the cosine super-resolver, but look at the enhancement of the secondaries in η, compared to those unaltered in ξ.

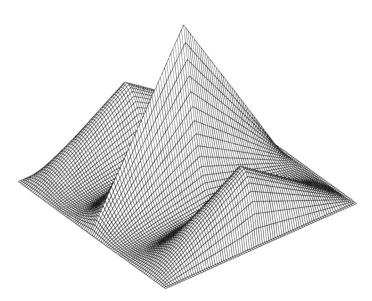

Figure 8.16 The *OTF* for the cosine super-resolver. Reminiscent of that for the double slit?

above, often a square telescope aperture, oriented so that its diagonal is parallel to a line connecting the two stars, will allow imaging one star in the diffraction "groove" at 45° to both ξ- and η-axes.

8.1.2.2 Double Aperture Super-Resolution It seems that any double aperture would be a super-resolver; the amplitude transmission is larger at the edges than in the center. At the end of Chapter 6 we convolved a double aperture, and Figure 8.16 was reminiscent of that. Here is an example somewhat familiar to many, the fully illuminated double-square's diffraction patterns. Figure 8.17(a) is $E(\xi, \eta)$ and Figure 8.17(b) is the irradiance function. Notice the skinny central maximum caused by the center-to-center spacing of the two squares and the increased secondaries which we usually call interference fringes. You can tell that these were two squares because the diffraction envelope over the interference fringes is the same in both ξ and η. That broad central area represents the extent of the central maximum in the diffraction pattern of just one square. The center-to-center spacing of the two squares was three times the width of one square, like an illustration in Hecht's *Optics* (1987).

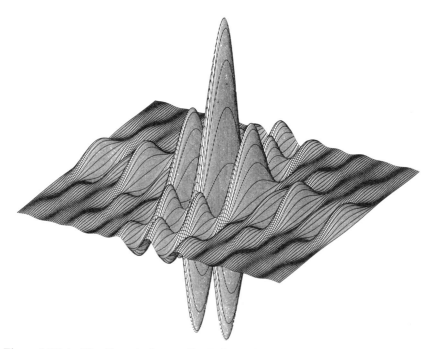

Figure 8.17(a) The Fraunhofer amplitude for a double-square aperture, the aperture of dimensions commonly used by textbook authors; but usually printed is only a two-dimensional slice of Figure 8.17(b). SURFER helps reveal so much more in the double-slit pattern.

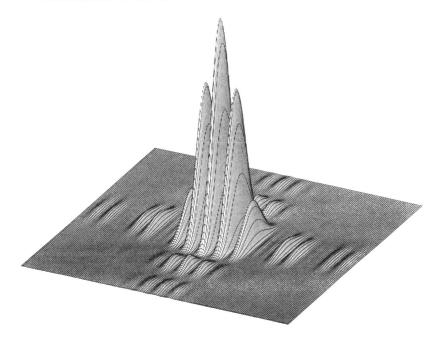

Figure 8.17(b) The Fraunhofer irradiance for a double-square aperture, both squares fully illuminated. The center-to-center spacing of the two squares was thrice the width of one square.

In the previous paragraph I said "fully illuminated" so as not to cheat the reader as is often done in a physics teaching lab or in books. In books the interference pattern, however it was made, is usually shown only along the axis of the interference fringes; the photographs are usually cropped. Even in the lab, where narrow long slits are illuminated with an unspread laser beam, the fringes show up nicely in ξ, but in η the irradiance should decrease in the manner of the laser beam, i.e., a Gaussian decrease if the full length of the slits are *not* illuminated. Figures 8.17(a) and 8.17(b) were easily created in SURFER.

8.2 ILLUSTRATING THE SHIFT THEOREM IN SIMPLE DIFFRACTION: PHASE IN A FOURIER TRANSFORM

Let us illustrate the rather surprising result that comes about if the aperture is translated in its plane. We will shift the square aperture, $A(x, y)$, a distance $+s$ along the x-axis. Though our shifted aperture can be described as $A(x - s, y) = 1$, we will let the aperture be defined by the integration limits.

$$E(\xi, \eta) = \int_{-a/2}^{+a/2} \int_{-a/2+s}^{+a/2+s} e^{-i2\pi(\xi x + \eta y)} \, dx \, dy$$

APODIZATION AND SUPER-RESOLUTION

We will more quickly get to the results if we do this integral in a slightly different form,

$$E(\xi, \eta) = \int_{-a/2}^{+a/2} e^{-i2\pi\eta y} \, dy \cdot \left[\int_{-a/2+s}^{0} e^{-i2\pi\xi x} \, dx + \int_{0}^{+a/2+s} e^{-i2\pi\xi x} \, dx \right]$$

The first integral in ηy we have done before; it yields $a \operatorname{sinc}(a\pi\eta)$. So we will have

$$E(\xi, \eta) = a \operatorname{sinc}(a\pi\eta) \left(\frac{-1}{i2\pi\xi} \right) \left[e^{-i2\pi\xi x} \Big|_{-a/2+s}^{0} + e^{-i2\pi\xi x} \Big|_{0}^{a/2+s} \right]$$

$$E(\xi, \eta) = a \operatorname{sinc}(a\pi\eta) \left(\frac{-1}{i2\pi\xi} \right) [e^{-i2\pi\xi \cdot 0} - e^{-i2\pi\xi(-a/2+s)}$$
$$+ e^{-i2\pi\xi(+a/2+s)} - e^{-i2\pi\xi \cdot 0}]$$

$$E(\xi, \eta) = a \operatorname{sinc}(a\pi\eta) \left(\frac{-1}{i2\pi\xi} \right) [1 - e^{+i2\pi\xi a/2} \cdot e^{-i2\pi\xi s} + e^{-i2\pi\xi a/2} \cdot e^{-i2\pi\xi s} - 1]$$

The ones add to zero; we factor out one exponential and reorder the exponentials with the -1 factor:

$$E(\xi, \eta) = a \operatorname{sinc}(a\pi\eta) \left(\frac{+1}{i2\pi\xi} \right) e^{-i2\pi\xi s} [e^{+i2\pi\xi a/2} - e^{-i2\pi\xi a/2}]$$

Now cancel the exponential 2's, multiply by a/a, and rearrange.

$$E(\xi, \eta) = a \operatorname{sinc}(a\pi\eta) \cdot \frac{a}{\pi\xi a} \cdot e^{-i2\pi s \xi} \left[\frac{e^{+i\pi a \xi} - e^{-i\pi a \xi}}{2i} \right]$$

This we rewrite as

$$E(\xi, \eta) = a \operatorname{sinc}(a\pi\eta) \cdot a \cdot e^{-i2\pi s \xi} \cdot \frac{\sin(a\pi\xi)}{a\pi\xi}$$

Now put that into final recognizable form,

$$E(\xi, \eta) = (e^{-i2\pi s \xi}) \cdot a^2 \operatorname{sinc}(a\pi\xi) \operatorname{sinc}(a\pi\eta)$$

To calculate the irradiance in the Fraunhofer diffraction pattern of this shifted square aperture we square E. But since E is complex (note the presence of

$i = \sqrt{-1}$), instead of calculating $I = E \times E$ we calculate $I = E \times E^*$, where $E^* = $ complex conjugate of E. That is:

$$I(\xi, \eta) = (e^{-i2\pi s\xi})(e^{+i2\pi s\xi})[a^2 \operatorname{sinc}(a\pi\xi) \operatorname{sinc}(a\pi\eta)]^2$$
$$I(\xi, \eta) = 1 \cdot [a^2 \operatorname{sinc}(a\pi\xi) \operatorname{sinc}(a\pi\eta)]^2$$
$$I(\xi, \eta) = a^4 \operatorname{sinc}^2(a\pi\xi) \cdot \operatorname{sinc}^2(a\pi\eta)$$

What! This is identical to the irradiance of the unshifted square aperture! Within our Fraunhofer approximation, it matters not, the amount of the shift nor its direction insofar as the visual or photographic appearance of the diffraction pattern is concerned. The pattern remains fixed in the transform plane, even while the aperture is in translation! This is a rather surprising result; under the Fraunhofer conditions and approximations it is true. The complex factor $\exp(-i2\pi s\xi)$ in $E(\xi, \eta)$ alters the phase of the pattern, but not the appearance; we will elaborate on this. This result is important for the understanding of optical correlation. A structure can be shifted to any part of the aperture but its diffraction pattern maintains the same appearance.

If an aperture has a Fraunhofer diffraction amplitude E_0, then the same aperture shifted a distance $+p$ on the y-axis will have a Fraunhofer amplitude $e^{-i2\pi p\eta} \cdot E_0$, a mathematically complex expression. We can refer to this as the Shift Theorem. Very often the Shift Theorem is stated in the following manner. If an aperture $A(x, y)$ is shifted $+s$ in the $+x$-direction, and $+p$ in the $+y$-direction, its new name becomes $A(x - s, y - p)$.

If $A(x, y)$ Fourier transforms to $E(\xi, \eta)$,
then $A(x - s, y - p)$ Fourier transforms to $(e^{-i2\pi s\xi})(e^{-i2\pi p\eta})E(\xi, \eta)$,
or $e^{-i2\pi(s\xi + p\eta)}E(\xi, \eta)$

The sign on these exponentials would be positive if your definition of the Fourier transform uses a positive exponential.

The phase factor is an important element in this Fourier transform; here it contains the position information for this single aperture. We know that the inverse transform will yield back the original aperture, and should we do the inverse transformation of just E_0 we will get back, as we know, the square aperture *centered* in the aperture plane. This Fourier transform, $E_0 \exp(-i2\pi s\xi)$, is not a discrete collection of spectral terms but a continuous function with the phase of each spatial frequency being a function of ξ, here a simple linear function of ξ. The phase function is only part of the exponential. We say the phase function is $\Phi(\xi, \eta) = -2\pi s\xi$.

The Shift Theorem and its inverse, the Modulation Theorem are summarized at the end of Chapter 9 which deals with complex apertures.

How shall we display phase? We ran into this problem earlier in Chapters 3 and 4 with one-dimensional "aperture" functions, and there we were able

228 APODIZATION AND SUPER-RESOLUTION

to display phase as a rotation, the helical display, about the single spatial frequency axis. But now in 3-space two axes are used for spatial frequencies ξ, η, and the third axis displays $E(\xi, \eta)$. This will require that phase be shown in a separate diagram.

In a proper optical setup, viewing the diffraction pattern in the Fourier plane, the focal plane of a lens, Mother Nature has taken care that the phase function is present in addition to irradiance. Even though one cannot see evidence of phase it is definitely an essential component of the Fourier transform. If the light is to be processed further, after the Fourier plane, phase may need to be carefully handled; it is sensitive over distances less than λ. Here is what this phase function, $\Phi(\xi, \eta) = -2\pi s\xi$, looks like (Figure 8.18) for $s = 0.001$ meter.

For a $+s$ shift in x and a $-p$ shift in y the phase function is $\Phi(\xi, \eta) = -2\pi s\xi + 2\pi p\eta$ (Figure 8.19), with $|s| = |p| = 0.001$ meter. We can expect with some other aperture functions, that the phase function may be nonlinear, nonplanar, for example, spherical.

It makes physical sense that if an aperture is shifted, some light reaching the diffraction plane will travel different distances, some more, some less, thus phase shifting the diffracted light waves relative to the unshifted aperture's diffracted light.

Doing the inverse transformation, i.e., reimaging the shifted aperture, will show it shifted, though the diffraction pattern was not space shifted, but phase shifted. If there was some way to destroy this phase function in the transform plane, force it everywhere to zero, then the reimaged aperture would not appear

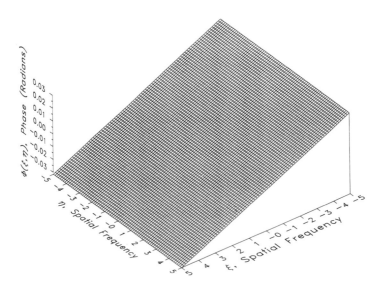

Figure 8.18 A two-dimensional linear phase function, $\Phi(\xi, \eta) = -2\pi s\xi$, with $s = 0.001$ meter.

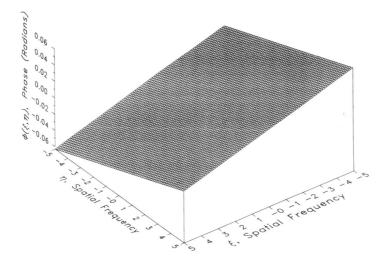

Figure 8.19 A two-dimensional phase function, linear in both ξ and η, $\Phi(\xi, \eta) = -2\pi s\xi + 2\pi p\eta$, with $|s| = |p| = 0.001$ meter.

physically shifted. How could one eliminate this shift-induced phase function? Conceptually, how in the transform plane could one eliminate our simple linear phase function? (A wedge of glass?)

8.3 MULTIPLE APERTURES: APPLICATION OF SHIFT AND LINEARITY PRINCIPLES

We have illustrated that the diffraction pattern of a shifted aperture "appears" to be identical to that of the unshifted aperture, but we know that the total amplitude of the wave disturbance creating the diffraction pattern is linearly phase-shifted parallel to the shift direction. The Fourier transform contains a phase factor, but "squaring" the transform (to get the irradiance) removes evidence of the phase factor.

Then consider the case of two square apertures, each of dimensions a by a and separated along the x-axis also by the distance a. That is, each by itself represents what we would term a shifted aperture, one shifted $s = +a$, the other $s = -a$. *Each by itself*, shifted, would produce a diffraction pattern identical "in appearance" to the diffraction pattern of the unshifted aperture, but we know that $E(\xi, \eta)$ for each will contain a phase-shift factor, one $\exp(-i2\pi a\xi)$ and the other, $\exp(+i2\pi a\xi)$. The reader is probably well aware that the diffraction pattern for these two square apertures will be that of a two-aperture interference pattern, two squares instead of a "double slit." We are dealing here with symmetrical shifts.

An interesting question is: knowing the two-aperture separation, i.e., "the shifts," and knowing the diffraction amplitude $E(\xi, \eta)$ of one unshifted square aperture, can we mathematically formulate the $E(\xi, \eta)$ and hence $I(\xi, \eta)$ for the double-square aperture? Yes we can, using the linearity principle of Fourier transforms. If aperture A has Fourier transform α and if aperture B has Fourier transform β, then aperture $(A + B)$ has Fourier transform $(\alpha + \beta)$.

For the "+" shifted aperture:

$$E_+(\xi, \eta) = (e^{-i2\pi a \xi})[a^2 \, \text{sinc}(a\pi\xi) \, \text{sinc}(a\pi\eta)], \qquad \text{mathematically "complex."}$$

For the "−" shifted aperture:

$$E_-(\xi, \eta) = (e^{+i2\pi a \xi})[a^2 \, \text{sinc}(a\pi\xi) \, \text{sinc}(a\pi\xi)], \qquad \text{mathematically "complex."}$$

We add these two amplitudes (the linearity principle) to obtain the $E(\xi, \eta)$ amplitude of the double aperture:

$$E = E_+ + E_- = (e^{-i2\pi a \xi} + e^{+i2\pi a \xi})[a^2 \, \text{sinc}(a\pi\xi) \, \text{sinc}(a\pi\eta)]$$

The Euler identity (which we don't use too often):

$$\cos \theta = \frac{(e^{i\theta} + e^{-i\theta})}{2}$$

allows us to write:

$$E(\xi, \eta) = 2 \cos(2\pi a \xi)[a^2 \, \text{sinc}(a\pi\xi) \, \text{sinc}(a\pi\eta)] \qquad (8.1)$$

This is mathematically "real." Within the square brackets is the expression for a single aperture's amplitude pattern. The irradiance is

$$E^2(\xi, \eta) = I(\xi, \eta) = 4[a^4 \, \text{sinc}^2(a\pi\xi) \, \text{sinc}^2(a\pi\eta)] \cos^2(2\pi a \xi) \qquad (8.2)$$

$E(\xi, \eta)$, $I(\xi, \eta)$, and $\log(1 + |E|)$ are plotted in Figures 8.20, 8.21, and 8.22, where $a = 1$. Within the square brackets of (8.2) is the irradiance expression for one aperture. The Fraunhofer irradiance for the symmetrical double aperture is this bracketted expression modulated by a cosine-squared function (interference fringes); or, if you prefer, cosine-squared fringes modulated by the diffraction pattern (bracketted expression) of one aperture; and it is four times brighter at the central peak than the single aperture. In this specific example the cosine-squared variations occur twice as frequently as the $[\text{sinc}(a\pi\xi)]^2$ variations.

Note in the amplitude expression (8.1) that if $2a = p =$ center-to-center separation of the two apertures, the $2 \cdot \cos(2\pi a \xi)$ factor becomes simply

MULTIPLE APERTURES 231

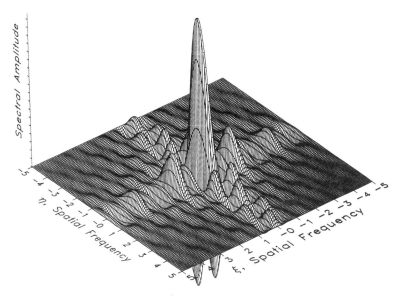

Figure 8.20 The Fraunhofer amplitude, $E(\xi, \eta)$, the Fourier transform, for two square apertures, the opaque center being the same width as each square. See Figure 8.23.

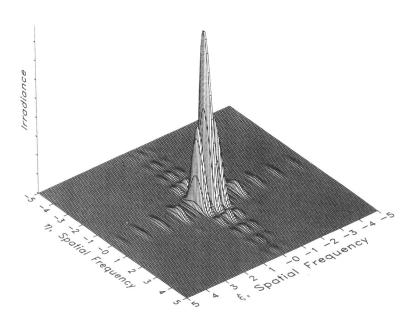

Figure 8.21 The Fraunhofer irradiance, $I(\xi, \eta)$, the square of the Fourier transform, for the double aperture of Figure 8.23.

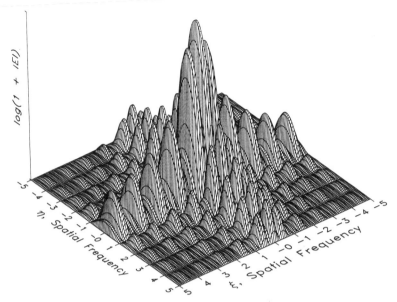

Figure 8.22 To enhance the secondaries of Figure 8.20, this is a plot of $\log(1 + |E(\xi, \eta)|)$.

$2 \cdot \cos(p\pi\xi)$. **Important conclusion:** If E_0 is the Fourier transform of any aperture, then a doubled aperture of center-to-center separation distance p in, say, the x-direction, will have a Fourier transform

$$2 \cdot \cos(p\pi\xi) \cdot E_0$$

This simple form results from the simple addition of Fourier transforms and using the Euler identity. $E(\xi, \eta)$ for this double aperture can also be obtained by subtracting E_0 (for the centered a-by-a aperture) from the $E(\xi, \eta)$ for a $3a$-by-a aperture on the x-axis; the linearity principle (Figure 8.23). These are differently spaced squares than those which produced Figures 8.17(a) and 8.17(b).

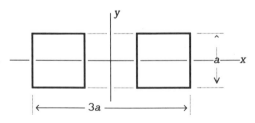

Figure 8.23 The double-square aperture with center-to-center spacing equal to twice the width of one square.

The superposition of the phase-shifted wave contributions from each of the shifted apertures has caused total cancellation of the disturbance in the diffraction plane at points where the cosine-squared minima occur. That is, at those points the total contribution to the amplitude of "+" waves is exactly matched by the total contribution of "−" waves. We find it rather curious that nature can occasionally balance things out so perfectly.

Although the central peak, $I(0, 0)$, has four times the radiant power of the pattern of the single aperture, the *total* pattern contains only twice the power let through a single aperture. (Two apertures let through twice the energy of one.) The cosine-squared factor dictates quite a different power distribution than that of the single aperture.

Well then, if we know $E(\xi, \eta)$ for two apertures shifted apart in the x-direction, then why not add two more identical apertures shifted in the y-direction from the first pair? Following the same procedures, and if we shift a distance $\pm a$ in the y-direction, we will simply create another factor $2 \cdot \cos(2\pi a\eta)$, or $2 \cdot \cos(p\pi\eta)$ and the irradiance will have cosine-squared fringes in the Y-direction also. Thus we have the diffraction pattern for the array of four identical apertures.

If

$$E_0(\xi, \eta) = a^2 \operatorname{sinc}(a\pi\xi) \operatorname{sinc}(a\pi\eta)$$

for the single optically centered a-by-a square aperture, then $E(\xi, \eta)$ for the aperture, double on the x-axis is

$$2 \cdot \cos(p\pi\xi) \cdot E_0(\xi, \eta)$$

Doubling this aperture the same way in the y-direction gives

$$E(\xi, \eta) = 2 \cdot \cos(p\pi\eta) \cdot 2 \cdot \cos(p\pi\xi) \cdot E_0(\xi, \eta)$$

From the linearity relationship this is the same as subtracting the Fourier transform of a cross from the Fourier transform of a 3a-by-3a square. If you try this be careful not to remove the center section twice (Figure 8.24). The duplication process described above can be carried on as long as is practical, e.g., to multiple apertures of 8, 16, 32, etc., individual openings. And because of the Linearity Theorem, other Fourier transforms can be added in, to handle odd numbers of openings.

Figures 8.25, 8.26, 8.27, 8.28, show, respectively: the symmetric four-squares aperture; $E(\xi, \eta)$, its Fourier transform; $|E(\xi, \eta)|$; and the aperture OTF obtained by convolution.

Figures 8.29 and 8.30 show a more widely spaced set of four square apertures and their convolution, their OTF.

Figures 8.31, 8.32, and 8.33 show, respectively: the symmetric four-circles aperture, but spaced a bit closer than the squares; its convolution, the OTF; and the irradiance pattern from such a four-hole aperture, showing what one would expect it to show.

234   APODIZATION AND SUPER-RESOLUTION

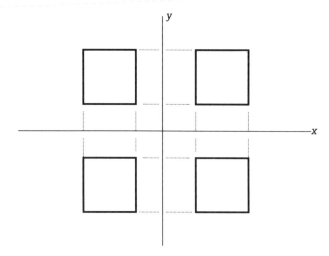

Figure 8.24   We shift the double-square aperture to form a symmetrical aperture of four squares. The "shift" concept is quite useful.

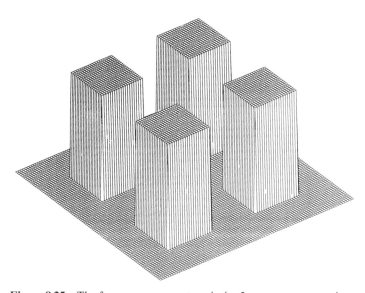

Figure 8.25   The four-squares aperture in its 3-space representation.

MULTIPLE APERTURES 235

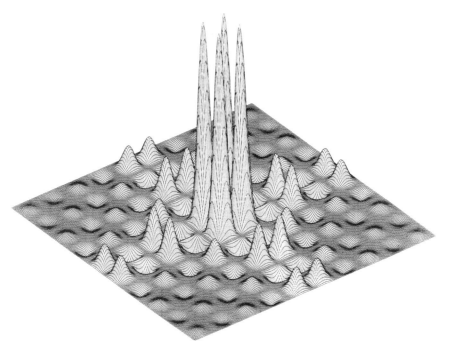

Figure 8.26 The Fourier transform of the function of Figure 8.25.

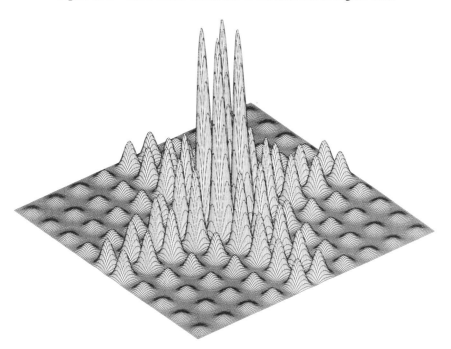

Figure 8.27 The absolute value of the function of Figure 8.26.

236 APODIZATION AND SUPER-RESOLUTION

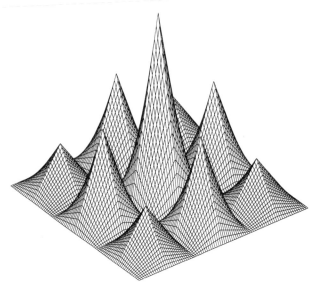

Figure 8.28 The self-convolution of the function of Figure 8.25, the *OTF*.

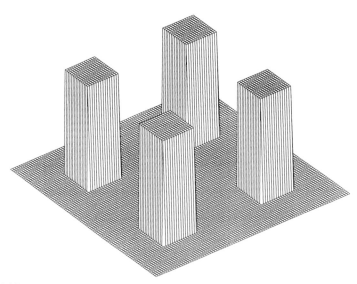

Figure 8.29 A more widely spaced but still symmetrical set of four square apertures.

MULTIPLE APERTURES 237

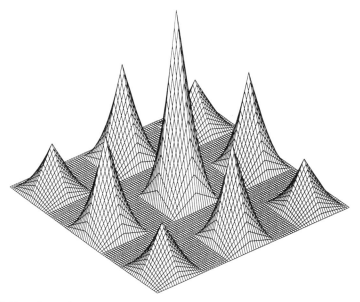

Figure 8.30 The self-convolution of the function of Figure 8.29, the *OTF*. You were undoubtedly expecting those flat regions.

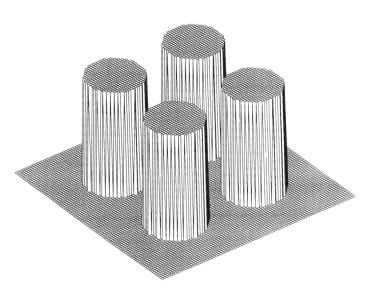

Figure 8.31 These symmetric four circular apertures are spaced a bit closer than their diameters; that should show up in the self-convolution.

238 APODIZATION AND SUPER-RESOLUTION

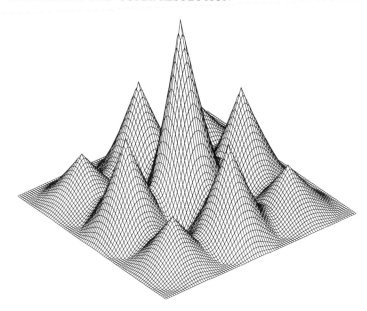

Figure 8.32 There are no zeros in the interior of this self-convolution of the function of Figure 8.31.

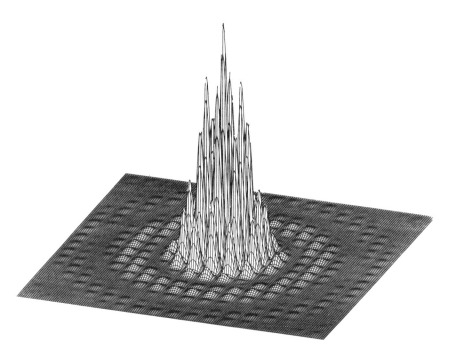

Figure 8.33 The Fraunhofer irradiance for the closely spaced aperture of four circles. There will be some limitations plotting in 3-space onto "A" size paper.

MULTIPLE APERTURES 239

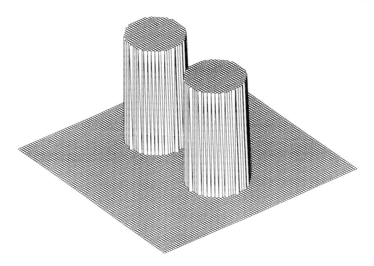

Figure 8.34 A close-spaced two-circle aperture.

Figures 8.34 and 8.35 are the close-spaced double-circular aperture and its convolution.

Related to Section 8.1, Figures 8.36 and 8.37 are a 50% ring aperture and its convolution.

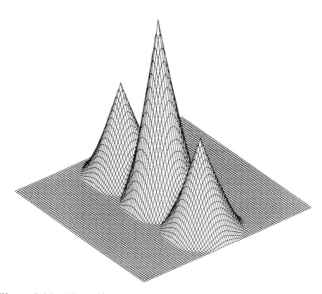

Figure 8.35 The self-convolution of the function of Figure 8.34.

240 APODIZATION AND SUPER-RESOLUTION

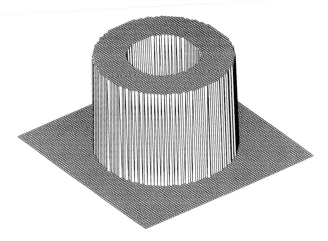

Figure 8.36 A circular ring aperture, only 50% radially blocked, the Fourier transform of which makes a nice calculation using the linearity principle.

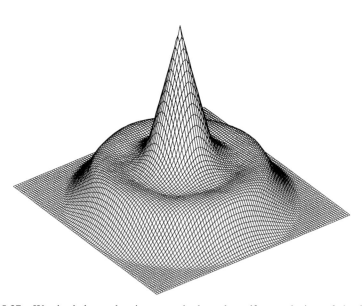

Figure 8.37 We don't have the time to calculate the self-convolution of the function of Figure 8.36; MATLAB did this one in a few minutes.

8.4 QUALITY APERTURES IN THE LABORATORY

Harrington and Winter (1984) have designed a set of apertures made by electron beam lithography in a chromium film on an optical flat. The plates are described in A. P. Harrington and A. T. Winter, "Aperture plates for Fraunhofer diffraction," *Eur. J. Phys.* (1984) pages 238–242. Figure 8.38 illustrates the apertures present on the optical flat. Commercial copies are likely to be available from W. A. Technology Ltd., Chesterton Mills, French's Road, Cambridge, England CB4 3NP, UK. We suggest contacting B. K. Ambrose, there, for the current price. In 1991 we paid £46. They may also be able to supply a reprint of the Harrington and Winter paper. Figure 8.38 is obviously a negative print.

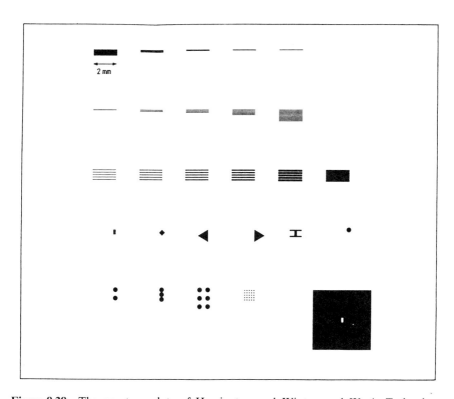

Figure 8.38 The aperture plate of Harrington and Winter, and W. A. Technology, England. See text. See their article for a nice optical arrangement. (A. P. Harrington and A. T. Winter, "Aperture plates for Fraunhofer diffraction," *Eur. J. Phys.* (1984), pp. 238–242. © The Institute of Physics & the European Physical Society. Reproduced with permission.)

QUESTIONS AND CHALLENGES TO PONDER

1. What is $E(0, 0)$ for a single unshifted square aperture?

2. What is $E(0, 0)$ for a single shifted square aperture?

3. Why is $E(0, 0)$ for the symmetric double-square aperture twice that of the single aperture?

4. Why is $I(0, 0)$ for the symmetric double-square aperture four times that of the single aperture?

5. You might attempt to calculate, from the irradiance expressions, the relative *total* power in the diffraction patterns of the double-square aperture and single-square aperture; or maybe just the relative power out to, say, ξ and η both equal to 5.

6. Calculate and plot the double-slit Fraunhofer diffraction pattern for slits fully illuminated in the x-direction (the spacing direction), but with a Gaussian decrease in the y-direction.

7. Calculate and plot the Fraunhofer diffraction pattern for a two-square aperture, but with squares of different sizes.

9

COMPLEX APERTURES

Up to this point we have only considered diffraction by apertures which were mathematically real. The aperture transmission function $A(x, y)$ described with real numbers the amplitude transmission at point (x, y) in the aperture. In Chapter 8 we apodized a square aperture using a cosine function to describe amplitude transmission. Since we are dealing with Fraunhofer diffraction, plane waves will be incident upon the aperture, thence diffracted by the aperture. Suppose instead of altering the *amplitude* of the disturbance as it goes through the aperture, we instead simply slow up, or delay, parts of the disturbance, i.e., alter the phase of certain parts.

9.1 PHASE CONCEPTS

For example, suppose we put a flat plate of glass with parallel faces in part of the aperture. It is transparent, ideally it need not alter the amplitude anywhere, but since light travels more slowly in glass than in air, the glass slows down that part of the disturbance passing through it. We say that that part of the disturbance will be shifted in phase, an amount depending upon the thickness of the glass. If d is the glass thickness, and n its index of refraction for the wavelength of light being used, then the product, nd, can be called the optical thickness and the phase shift $\Delta\phi = (2\pi/\lambda)nd$. Of course if nd is an integral multiple of λ, then no alteration of the diffraction pattern should occur. Let's test this idea but first let us consider how to mathematically introduce a phase shift into the aperture function $A(x, y)$.

COMPLEX APERTURES

A mathematical expression for a plane wave like ours traveling in time and space, for simplicity in the positive x-direction, is:

$$y(x, t) = y_0 \cos(\omega t - kx)$$

where $\omega = 2\pi f$ and $k = 2\pi/\lambda$.

$$\omega/k = 2\pi f/(2\pi/\lambda) = f\lambda = \lambda/T = \text{speed of the wave}$$

If the wave is phase shifted, say, advanced a distance d meters, of angular equivalent δ radians, i.e., moved in space and here specifically moved ahead one wavelength, $\delta = (2\pi/\lambda) \cdot d$, here $d = \lambda$, equivalent to 2π radians, then its expression would be:

$$y(x, t) = y_0 \cos(\omega t - kx - \delta) = y_0 \cos(\omega t - kx - 2\pi)$$

A sinusoidal wave shifted by 2π radians is identical to the unshifted wave. For a wave shifted in the positive direction, for the phase change we use $-(2\pi/\lambda)d$; in the negative direction we use $+(2\pi/\lambda)d$. In either case, if $d = \lambda$, the phase change is $\pm 2\pi$, resulting in a wave identical to the original wave. So if a glass plate in the aperture has an optical thickness, $nd = \lambda$, or an integral multiple of λ, there is no apparent difference in the diffraction with or without the plate. ($\lambda = \lambda$ in the glass.)

Using exponential notation (for later ease in mathematics) we will repeat the above calculation.

Since $e^{i\theta} = \cos\theta + i\sin\theta$, we say:

$$y(x, t) = \text{Real Part of } \{y_0 e^{i[\omega t - kx \pm (2\pi/\lambda)d]}\}$$

because $\cos\theta$ is the real part of $e^{i\theta}$; $\sin\theta$ is the imaginary part of $e^{i\theta}$. We shorten "Real Part of" to "Re," so that the expression becomes:

$$y(x, t) = \text{Re}\{y_0 e^{i[\omega t - kx \pm (2\pi/\lambda)d]}\}$$

but "everyone knows" what we mean by this notation so we shorten still further to:

$$y(x, t) = y_0 e^{i[\omega t - kx \pm (2\pi/\lambda)d]}$$

and it is "understood" that we mean only the real part of the expression. We rewrite that as

$$y(x, t) = y_0 e^{i\omega t} \cdot e^{-ikx} \cdot e^{\pm i(2\pi/\lambda)d} \qquad (9.1)$$

PHASE CONCEPTS   245

The part which shifts the phase, (delays or advances the wave) can be pulled out as a separate factor. The expression $e^{\pm i(2\pi/\lambda)d}$ is this phase-shift factor; perhaps $e^{\pm i(2\pi/\lambda)nd}$ would be more correct. We are unlikely to encounter devices which advance or speed up the wave so we will restrict our discussion to phase shifts which delay the wave:

$$e^{+i(2\pi/\lambda)nd} \text{ will be the phase shift factor for delay.}$$

If $nd = \lambda$, $e^{i(2\pi/\lambda)\lambda}$ represents a shift or delay of one wavelength, and this factor has a value of $e^{i2\pi}$; meaning

$$\text{Real Part of } e^{i2\pi} = \text{Re}\{\cos(2\pi) + i\sin(2\pi)\} = 1.$$

As we indicated earlier, this factor, a phase-shift factor (not function) of 1 in (9.1) changes nothing.

Therefore if our aperture function $A(x, y)$ contains something like glass of optical thickness nd, or perhaps a thickness which is variable over the points (x, y) of the aperture, we can adequately describe the phase delay with an aperture phase function $\phi(x, y)$, and a phase-shift factor

$$e^{+i(2\pi/\lambda)\phi(x,y)}$$

We use lower case phi, ϕ, for the aperture plane. The complete aperture function would then look something like

$$e^{+i(2\pi/\lambda)\phi(x,y)} \cdot A(x, y)$$

The dimensions of $\phi(x, y)$ are length, meters in MKS units, and $\phi(x, y)$ simply describes the optical thickness. ϕ and λ require identical units of measure. Multiplying ϕ by $(1/\lambda)$ converts the optical thickness to a pure number, a fractional part of λ, and multiplication by 2π converts that to a phase delay in radians. An optical thickness of λ (a $\phi(x, y)$ of λ), multiplied by $(1/\lambda)$ becomes a fraction equal to 1; multiplied by 2π, it becomes a 2π radian phase delay; 2π radians represents one complete cycle of the wave disturbance; as before, it means no change. In regions where $\phi(x, y) > 0$, the wave is delayed or retarded. $\phi(x, y) < 0$ seems to not be physically possible.

9.1.1   Half Aperture With π Phase Lag

What happens if one-half the aperture is covered with a $\lambda/2$ phase delay device while the other half remains unaltered? This is the integral to be solved; the phase factor will be $e^{i\pi}$ for positive x.

$$E(\xi, \eta) = \iint\limits_{\text{aperture}} A(x, y) e^{-i2\pi(\xi x + \eta y)} \, dx \, dy$$

$$E(\xi, \eta) = \int_{-a/2}^{+a/2} e^{-i2\pi\eta y} \, dy \left\{ \int_{-a/2}^{0} e^{-i2\pi\xi x} \, dx + \int_{0}^{+a/2} e^{i\pi} \cdot e^{-i2\pi\xi x} \, dx \right\}$$

The Separation Theorem allows us to write it this way. We can either solve this or use the linearity principle and results from our past solutions. Let's take that second route. We'll treat this as two half apertures, each $a/2$ wide in x, by a wide in y, one shifted $+a/4$, the other shifted $-a/4$, along the x-axis. The one plus-shifted has the $\lambda/2$ or π phase delay. That can be seen in the above integral and in Figure 9.1 where, as usual, the all positive quadrant is in the foreground.

By inspection of such half-wide apertures we write

$$E(\xi, \eta) = a\,\text{sinc}(\pi a\eta) \cdot \frac{a}{2} \text{sinc}\left(\pi \frac{a}{2} \xi\right) \cdot \underbrace{e^{+i2\pi(a/4)\xi}}_{\text{minus shift}}$$

$$+ a\,\text{sinc}(\pi a\eta) \cdot \frac{a}{2} \text{sinc}\left(\pi \frac{a}{2} \xi\right) \cdot \underbrace{e^{-i2\pi(a/4)\xi}}_{\text{plus shift}} \cdot \underbrace{e^{i\pi}}_{\phi\,\text{lag}} \quad (9.2)$$

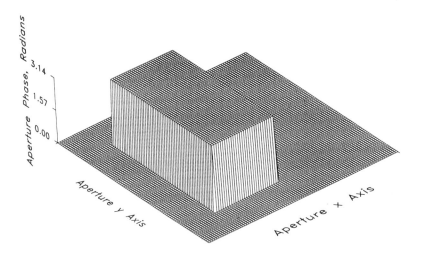

Figure 9.1 Half aperture with a π step in phase.

This is the sum of the Fourier transforms of two half-wide apertures side by side on the x-axis. The plus-shifted one has the phase lag. With $e^{i\pi} = -1$ this can be rearranged, and with multiplication by i/i it becomes

$$E(\xi, \eta) = i\left\{a^2 \operatorname{sinc}(\pi a \eta) \operatorname{sinc}\left(\pi \frac{a}{2}\xi\right) \sin\left(\frac{\pi}{2} a\xi\right)\right\}$$

This expression is purely imaginary. The amplitude of the spectral components is given in the expression within the brackets; for display it requires 3-space. The i multiplier means that all spectral components in $E(\xi, \eta)$, the Fourier transform, are shifted in phase $\pi/2$. $\Phi(\xi, \eta)$ (capital phi in the transform plane), the phase here is constant, $\pi/2$. For $a = 1$, Figure 9.2 shows this spectral amplitude. Figure 9.3 is the absolute value of the spectral amplitude, while Figure 9.4 shows the irradiance (the Fraunhofer diffraction pattern) for such an aperture. Do you see how nicely symmetrically split it is along the ξ-axis? The phase part of the aperture function, $\phi(x, y)$, was shown in Figure 9.1, while the real part of the aperture function looks like Figure 9.5, a simple square aperture.

9.1.2 Half Aperture With $\pi/2$ Phase Lag

If the square aperture has, in its plus-shifted half, a phase lag of only $\lambda/4$ or $\pi/2$ radians, then (9.2) becomes

$$E(\xi, \eta) = a \operatorname{sinc}(\pi a \eta) \cdot \frac{a}{2} \operatorname{sinc}\left(\pi \frac{a}{2}\xi\right) \cdot e^{+i2\pi(a/4)\xi}$$

$$+ a \operatorname{sinc}(\pi a \eta) \cdot \frac{a}{2} \operatorname{sinc}\left(\pi \frac{a}{2}\xi\right) \cdot e^{-i2\pi(a/4)\xi} \cdot e^{i\pi/2}$$

Now, $e^{i\pi/2} = \cos(\pi/2) + i \sin(\pi/2) = 0 + i = i$, thus,

$$E(\xi, \eta) = a \operatorname{sinc}(\pi a \eta) \cdot \frac{a}{2} \operatorname{sinc}\left(\pi \frac{a}{2}\xi\right)\{e^{+i\pi a\xi/2} + i e^{-i\pi a\xi/2}\}$$

Interpretation of that might seem a bit intimidating; let's just utilize the Euler relationship.

$$E(\xi, \eta) = a \operatorname{sinc}(\pi a \eta) \cdot \frac{a}{2} \operatorname{sinc}\left(\pi \frac{a}{2}\xi\right)\{\cos(\pi a\xi/2) + i \sin(\pi a\xi/2)$$

$$+ i[\cos(\pi a\xi/2) - i \sin(\pi a\xi/2)]\}$$

248 COMPLEX APERTURES

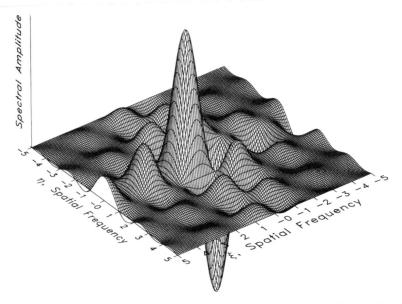

Figure 9.2 Amplitude of the spectral components in the Fourier transform of the phase function of Figure 9.1.

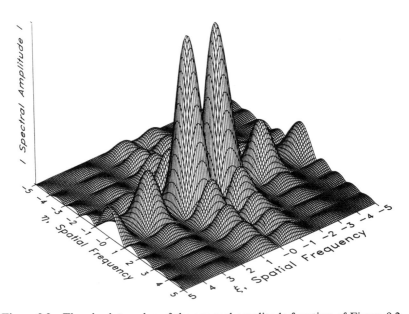

Figure 9.3 The absolute value of the spectral amplitude function of Figure 9.2.

PHASE CONCEPTS 249

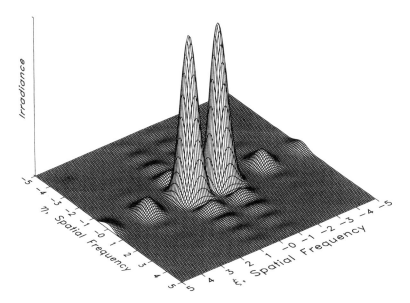

Figure 9.4 The Fraunhofer irradiance for the phase aperture of Figure 9.1.

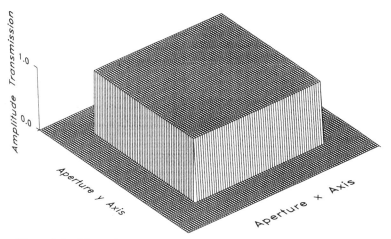

Figure 9.5 The real part of the phase aperture function of Figure 9.1.

Being careful with the i's, this becomes

$$E(\xi, \eta) = a\, \mathrm{sinc}(\pi a\eta) \cdot \frac{a}{2} \mathrm{sinc}\left(\pi \frac{a}{2} \xi\right) \{\cos(\pi a\xi/2) + \sin(\pi a\xi/2)$$
$$+ i[\cos(\pi a\xi/2) + \sin(\pi a\xi/2)]\}$$

250   COMPLEX APERTURES

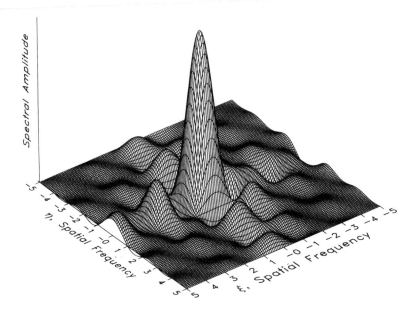

Figure 9.6 The $\pi/2$ phase half aperture has a complex Fourier transform with identical real and imaginary parts. This is the real part.

This function is not purely imaginary, but complex. The real and the imaginary parts are identical everywhere. A SURFER plot of the real part is Figure 9.6, while the absolute value is shown in Figure 9.7. Irradiance is Figure 9.8. The phase function in the Fourier transform plane (capital Φ)

$$\Phi(\xi, \eta) = \arctan(\text{Imaginary part of } E/\text{Real part of } E)$$
$$= \arctan(1) = \pi/4,$$

everywhere, since the real and imaginary parts are equal. There is a $\pi/4$ or $\lambda/8$ shift in phase for all spatial frequency spectral components. MATLAB can convolve this complex aperture with itself producing a complex OTF. The modulation transfer function, MTF, is Figure 9.9, the phase transfer function, PTF, is Figure 9.10.

9.1.3 Aperture With a Linear Phase Function

Suppose one could construct a complex aperture function which linearly varied the phase across an aperture a-wide in the manner shown in Figure 9.11. This will be a general case from which specific cases can be determined. We will need only the slope m of the linear aperture phase function and its ϕ intercept b at $y = 0$. For the general one-dimensional case shown $\phi(x, y) = my + b$, m can take $\pm$ values. Recall that when the square aperture was translated (shifted)

PHASE CONCEPTS 251

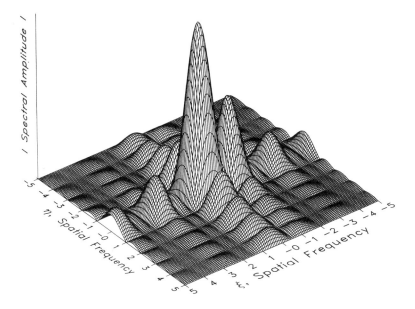

Figure 9.7 The absolute value of the function of Figure 9.6.

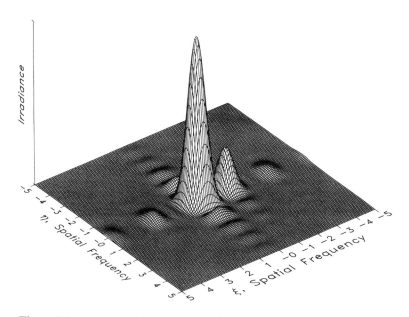

Figure 9.8 The Fraunhofer irradiance for the $\pi/2$ phase half aperture.

252 COMPLEX APERTURES

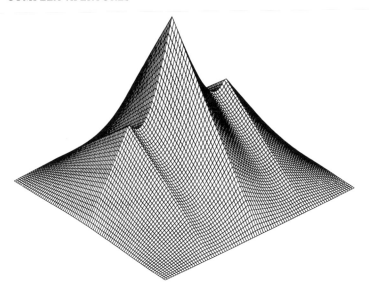

Figure 9.9 The modulation transfer function, MTF, for the $\pi/2$ phase half aperture.

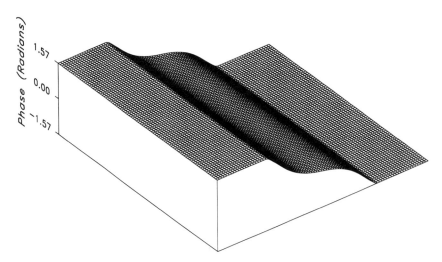

Figure 9.10 The phase transfer function, PTF, for the $\pi/2$ phase half aperture.

away from symmetry in the $x-y$ plane there resulted in the $\xi-\eta$ (or $X-Y$) plane a linear phase shift across the diffraction pattern. One might suspect some sort of reciprocity then in this example. An aperture phase function $\phi(x, y)$ with any negative values will lack physical realizability. It would mean advancing the phase of a light wave. How could you do that? For

PHASE CONCEPTS   253

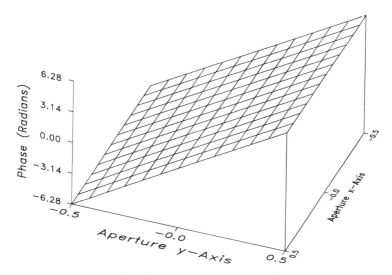

Figure 9.11 An aperture linear phase function.

this $\phi(x, y)$ linear phase,

$$E(\xi, \eta) = \int_{-a/2}^{+a/2} \int_{-a/2}^{+a/2} e^{+i(2\pi/\lambda)(my+b)} \cdot e^{-i2\pi(\xi x + \eta y)} \, dx \, dy$$

$$E(\xi, \eta) = \int_{-a/2}^{+a/2} e^{-i2\pi\xi x} \, dx \int_{a/2}^{+a/2} e^{i2\pi b/\lambda} \cdot e^{-i2\pi(\eta - m/\lambda)y} \, dy$$

The first integral we know. Rearrange the second,

$$E(\xi, \eta) = a \, \text{sinc}(\pi a \xi) e^{+i2\pi b/\lambda} \int_{-a/2}^{+a/2} e^{-i2\pi(\eta - m/\lambda)y} \, dy$$

Similar to our other solutions, this becomes

$$E(\xi, \eta) = a^2 e^{+i2\pi b/\lambda} \, \text{sinc}(\pi a \xi) \, \text{sinc}[\pi a(\eta - m/\lambda)]$$

In a specific example, if $m = +2\lambda/a$, $b = \lambda$, $a = 1$, then

$$E(\xi, \eta) = a^2 e^{+i2\pi} \, \text{sinc}(\pi \xi) \, \text{sinc}(\pi a \eta - 2\pi)$$

There is a uniform 2π phase shift over the entire $E(\xi, \eta)$ pattern and, as we suspected, the linear phase shift in y in the aperture has translated the diffraction pattern in η. More specifically, the aperture phase shift, $\phi(x, y)$, increasing in the $+y$-direction with a maximum delay of 2λ, i.e., 4π radians (one cycle), has caused the Fourier spectrum to be shifted in the $+\eta$-direction (or $+Y$-direction),

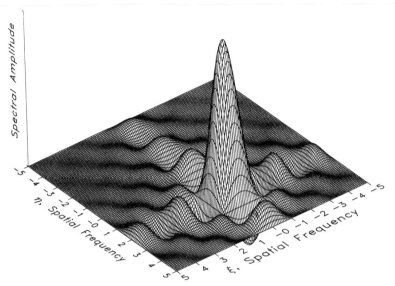

Figure 9.12 An aperture with a linear phase variation will have a spatially shifted Fraunhofer diffraction pattern. This is the real part of the Fraunhofer amplitude.

an amount in η of 2π radians. Figures 9.12 and 9.13 show $E(\xi, \eta)$ and $|E(\xi, \eta)|$. Apparently there will be no odd $\phi(x, y)$ functions; we cannot advance the phase of light. We can refer to this result as the Modulation Theorem; it and others are summarized at the end of this chapter. Thus we have a general approach to linear phase variation in an aperture.

9.1.4 Aperture V-Linear Phase Function

We now wish to present an example which appears to be simple but which can easily lead one astray. The aperture is again *a*-by-*a* but now with a V-shaped linear phase function in the *x*-direction. The maximum phase delay due to the $\phi(x, y)$ will eventually be set at $\lambda/2$ or π radians. The description of the aperture function is Figure 9.14; it has the appearance of the inverse of a Fresnel biprism. Since there is no variation in *y* by the Separation Theorem we can write part of $E(\xi, \eta)$ by inspection; the rest to be worked out.

$$E(\xi, \eta) = a\,\text{sinc}(\pi a \eta) \left\{ \int_{-a/2}^{0} e^{+i(2\pi/\lambda)(-\lambda x/a)} e^{-i2\pi \xi x}\, dx \right.$$
$$\left. + \int_{0}^{+a/2} e^{+i(2\pi/\lambda)(+\lambda x/a)} e^{-i2\pi \xi x}\, dx \right\}$$

PHASE CONCEPTS   255

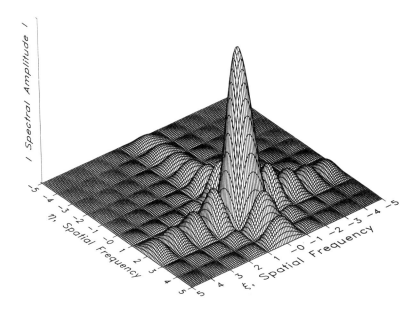

Figure 9.13 The absolute value of the Fraunhofer amplitude of Figure 9.12.

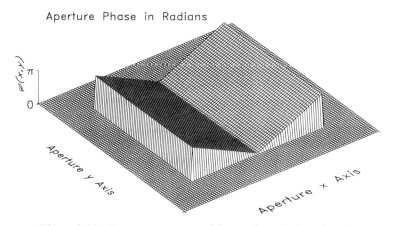

Figure 9.14 A square aperture with a V-shaped phase function.

At $x = 0$, the phase factor is $e^0 = 1$; at $x = \pm a/2$, the phase factor is $e^{+i\pi}$. Although the solution is not difficult it is something of a problem to put it in an order one can recognize. So instead of brute force solution of the above we'll use the results already formulated. We'll treat it as two half apertures: one $a/2$ wide with negative slope phase, shifted $-a/4$ in x; the other $a/2$ wide with positive slope phase, shifted $+a/4$ in x. Then we'll get the

256 COMPLEX APERTURES

Fourier transform of each separately and sum them. For the left and right halves we'll have

$$E(\xi, \eta)_{\text{left}} = a\,\text{sinc}(\pi a \eta)\,\underbrace{\frac{a}{2} e^{+i2\pi b/\lambda} \text{sinc}[\pi a(\xi - m/\lambda)]}_{\text{centered half aperture with phase function}}\,\underbrace{e^{+i2\pi a\xi/4}}_{\substack{-a/4 \\ \text{shift}}}$$

$$E(\xi, \eta)_{\text{right}} = a\,\text{sinc}(\pi a \eta)\,\underbrace{\frac{a}{2} e^{+i2\pi b/\lambda} \text{sinc}[\pi a(\xi - m/\lambda)]}_{\text{centered half aperture with phase function}}\,\underbrace{e^{-i2\pi a\xi/4}}_{\substack{+a/4 \\ \text{shift}}}$$

Now in both of these b is the $\phi(x, y)$ intercept at the center of the centered half aperture, $b = \lambda/4$. The slope of the aperture phase function is $m = -\lambda/a$ on the left, $+\lambda/a$ on the right. Letting $a = 1$, disregarding numerical constants, remembering that $\exp(i\pi/2) = 1$, and using the Euler relationship,

$$E(\xi, \eta)_{\text{left}} = -\text{sinc}(\pi\eta)\cdot\text{sinc}(\pi\xi/2 + \pi/2)\cdot\sin(\pi\xi/2)$$
$$+ i[\text{sinc}(\pi\eta)\cdot\text{sinc}(\pi\xi/2 + \pi/2)\cdot\cos(\pi\xi/2)]$$

$$E(\xi, \eta)_{\text{right}} = +\text{sinc}(\pi\eta)\cdot\text{sinc}(\pi\xi/2 - \pi/2)\cdot\sin(\pi\xi/2)$$
$$+ i[\text{sinc}(\pi\eta)\cdot\text{sinc}(\pi\xi/2 - \pi/2)\cdot\cos(\pi\xi/2)]$$

The sum of these two will be the Fourier transform of this (seemingly simple) complex aperture function.

$$E(\text{left}) + E(\text{right}) =$$
$$E(\xi, \eta) = \text{sinc}(\pi\eta)\cdot\sin(\pi\xi/2)[\text{sinc}(\pi\xi/2 - \pi/2) - \text{sinc}(\pi\xi/2 + \pi/2)]$$
$$+ i\{\text{sinc}(\pi\eta)\cdot\cos(\pi\xi/2)[\text{sinc}(\pi\xi/2 - \pi/2) + \text{sinc}(\pi\xi/2 + \pi/2)]\}$$

This Fourier transform is complex. SURFER can plot the real part, as shown in Figure 9.15; the imaginary part is shown in Figure 9.16. In SURFER's Grid menu we can form the modulus, which shows the amplitude of each spatial frequency spectral component,

$$\text{modulus of } E(\xi, \eta) = \sqrt{[\text{Re}\{E(\xi, \eta)\}]^2 + [\text{Im}\{E(\xi, \eta)\}]^2}$$

see Figure 9.17; and the phase in the Fourier transform,

$$\Phi(\xi, \eta) = \arctan\frac{\text{Im } E(\xi, \eta)}{\text{Re } E(\xi, \eta)}$$

see Figure 9.18. Figure 9.19 shows the topography of the modulus.

PHASE CONCEPTS 257

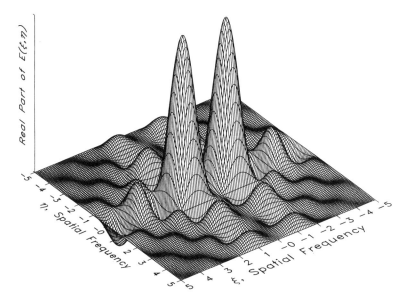

Figure 9.15 The real part of the Fourier transform of the aperture function of Figure 9.14.

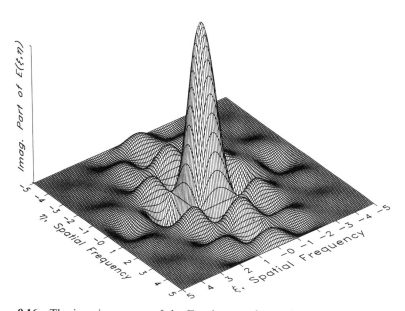

Figure 9.16 The imaginary part of the Fourier transform of the aperture function of Figure 9.14.

258   COMPLEX APERTURES

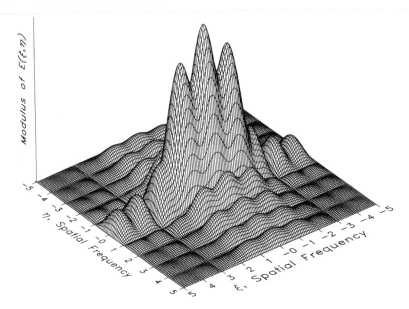

Figure 9.17 The modulus of the complex Fourier transform of the aperture function of Figure 9.14. Modulus $= \sqrt{(\text{Re } E)^2 + (\text{Im } E)^2}$.

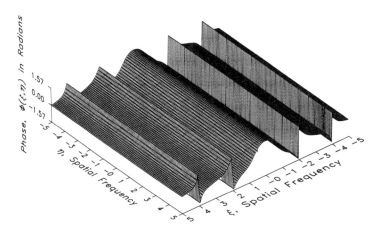

Figure 9.18 The phase of the complex Fourier transform of the aperture function of Figure 9.14. Phase $= \arctan(\text{Im } E/\text{Re } E)$. Done in SURFER.

Because of the symmetry here, all constants for all terms were the same; we ignored them. One needs to be cautious about that, for had the constants not been the same, the imaginary part and the real part would have been weighted differently in evaluations of the transform modulus and phase.

This aperture can be established in MATLAB and the above illustrations created there. The MATLAB convolution of the complex aperture function

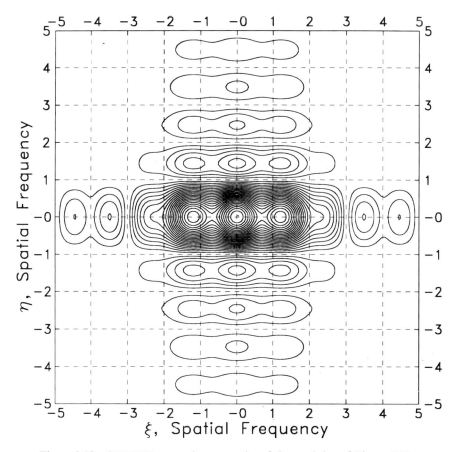

Figure 9.19 SURFER-created topography of the modulus of Figure 9.17.

yields a complex *OTF*. The *MTF*, the absolute value of the generated convolution, is Figure 9.20 and the *PTF* is Figure 9.21. The *PTF* was obtained with the angle command and unwrapped a few times; the phase steps in *PTF* are π.

Here are the MATLAB commands:

```
[x,y]=meshdom(-1:2/19:1, -1:2/19 : 1);
vphase=abs(0.5.*x);
ap=zeros(90);
ap(23:67,23:67)=ones(45).*exp(i*2*pi*vphase);
```

The command mesh(ap) will show only the real part, just the square aperture. The command mesh(angle(ap)) will show the vphase function. The commands we have illustrated before will yield the transforms and the *OTF*, its modulus and its phase, it being complex.

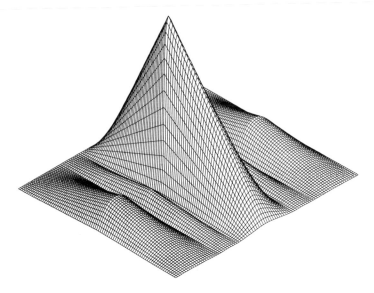

Figure 9.20 The modulation transfer function, *MTF*, of the complex optical transfer function, *OTF*, of the aperture of Figure 9.14. Created by MATLAB.

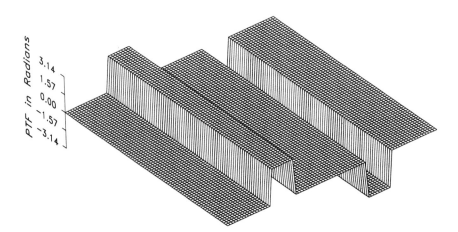

Figure 9.21 The phase transfer function, *PTF*, of the complex optical transfer function. *OTF*, of the aperture of Figure 9.14. Created by MATLAB.

9.2 MORE COMPLEX APERTURE FUNCTIONS; ABERRATIONS

Now into the realm of the more difficult: visualization of some primary aberrations. We will be examining the *PSF* in the focal plane of a lens; the lens function is to create a spherical wave front convergent into a point on the optical axis in the focal plane, i.e., convergent into the focal point. The

MORE COMPLEX APERTURE FUNCTIONS; ABERRATIONS   261

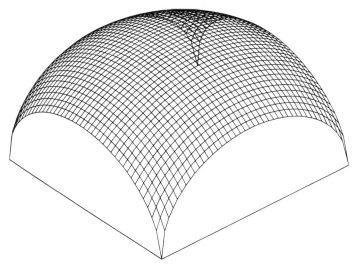

Figure 9.22 The representation of a converging spherical wave front. The surface represents the convergent phase surface.

lens will not do this perfectly; there will be deviations from a perfectly spherically convergent wave front. It is these deviations from spherical which constitute what are called aberrations. Figure 9.22 can represent a spherical wave front convergent into its center of curvature, while Figure 9.23 is meant to illustrate a deviation from perfectly spherical. You can see that it has no single center of curvature. It is an aberrated wave front. These aberrations are

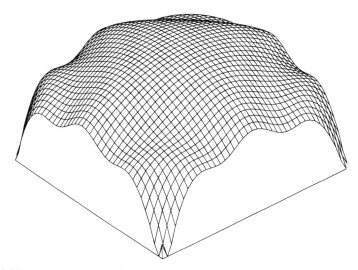

Figure 9.23 Representation of a converging wave front with some deviations from perfectly spherical, i.e., with some aberrations. It is the convergent phase surface.

262 COMPLEX APERTURES

described in the *exit pupil* of the optical system which can often be taken as the "plane" of the lens in a single-lens system. These deviations are not amplitude deviations but phase deviations; they are complex functions and we now know how to handle them.

Coma, in a circular aperture, is the first we will tackle. This aperture function will be complex, but from what has just gone before, we can do it. We are guided here by Kim and Shannon (1987). The integral we need to solve is the following one, where the complex aberration function is $V(x, y)$, in Cartesian coordinates. We see the aberration function making its appearance in exactly the same manner we described a more general phase function in the aperture plane; we called it $\phi(x, y)$. The general phase factor in the integral was written $e^{i(2\pi/\lambda)\phi(x,y)}$ and ϕ had dimensions of length which when multiplied by $1/\lambda$ resulted in a pure number. Now the aberration factor in the integral does not contain $1/\lambda$, so $V(x, y)$ must be described as a pure number, i.e., as a (variable) fraction of λ. $V(x, y)$ will describe the phase lag caused by the aperture in terms of fractions of λ, the wavelength of the light coming through.

$$E(\xi, \eta) = \iint_{\text{aperture}} e^{i2\pi V(x,y)} A(x, y) e^{-i2\pi(\xi x + \eta y)} \, dx \, dy$$

In all the cases we will examine the amplitude transmission function, $A(x, y) = 1$. If you consult Kim and Shannon you will see that they used a positive exponential, $e^{+i2\pi(\xi x + \eta y)}$ in this Fourier transformation; we remarked on this alternate definition earlier. Also Kim and Shannon include a factor before the integral equal to (1/aperture area); we have also remarked earlier about disregarding constant factors outside the integral, always making relative comparisons.

Kim an Shannon describe the aberration function in polar coordinates for what are termed the Seidel aberrations as

$$V(r, \theta) = a_{40} r^4 + a_{31} r^3 \cos \theta + a_{20} r^2 + a_{22} r^2 \cos^2 \theta + a_{11} r \cos \theta$$

These terms of this polynomial describe, successively, spherical aberration, coma, astigmatism, field curvature and distortion. Higher-order terms can be added in and $V(r, \theta)$ becomes rapidly complex.

But there is a simpler way to describe the aberrations, due to Frits Zernicke, using a set of orthogonal circular polynomials normalized for within a circle of unit radius. Kim and Shannon list thirty-seven Zernicke polynomials in polar coordinates, R and θ, in the imaging system exit pupil.

9.2.1 Coma

We will use Zernicke Polynomial Number 7 for coma in the y-direction, given by

$$\sqrt{8}(3R^2 - 2)R \cos \theta$$

MORE COMPLEX APERTURE FUNCTIONS; ABERRATIONS   263

Watch how this is coded for MATLAB in what follows; coma is the only one we will describe in detail. Kim and Shannon provide a "Graphic Library of Zernicke Polynomials," 37 of them, where they provide: topographic and isometric plots of the wave-front error for the 37 Zernicke polynomials, the isometric point spread function, and the isometric modulation transfer function. They do not display (as we shall) the phase transfer function, which, when it goes negative, describes those spatial frequencies at which there will be contrast reversal, i.e., object spatial frequencies that are white are imaged as black, and vice versa.

Coma done in MATLAB

In MATLAB:

```
[x,y]=meshdom(-1:2/89:1, -1:2/89:1);
r=sqrt(x.^2+y.^2);
theta=atan2(y,x);
```

These establish four matrices; all 90 × 90, one called x, one called y; one called r, one called theta.

```
coma=sqrt(8)*(3*(r.^2)-2).*r.*cos(theta);
```

This represents the phase deviation from perfectly spherical emerging from the aperture. The command mesh(coma) shows a phase variation in a 90 × 90 grid. We show some of these in low resolution as if seen on the computer screen (Figure 9.24(a)). Now put the phase variation into a round area, the exit pupil:

```
ap=zeros(90);
ap=circ(45).*exp(i*2*pi*(.2)*coma);
mesh(abs(ap))
```
will show a flat circle (Figure 9.24(b)), there is only phase variation in the pupil.

```
mesh(angle(ap))
```

will display the phase wave front (its deviation from the perfect sphere) emerging from the round pupil (Figure 9.24(c)). The command unwrap does not help. We can get rid of some of the spiking (round-off error) at the edge of the pupil by going to an 88 × 88 array and a 44 radius circle.

```
ap(2:89,2:89)=circ(44).*exp(i*2*pi*(.2)*coma(2:89,2:89));
mesh(angle(ap))
```

264   COMPLEX APERTURES

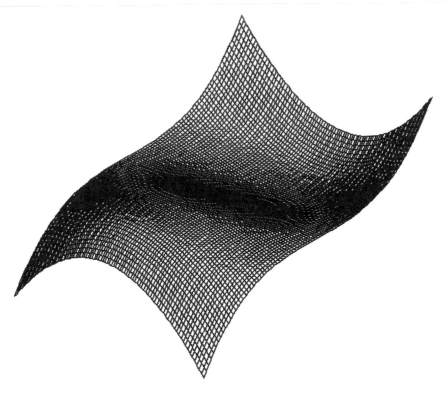

Figure 9.24(a)   The coma aberration function as viewed right off the screen.

should look somewhat better (Figure 9.24(d)). This is a good resolution, simple portrayal of a complex pupil function but it is too large for Fourier transforming. Make a smaller one:

```
[x,y]=meshdom(-1:2/19:1, -1:2/19:1);
r=sqrt(x.^2+y.^2);
theta=atan2(y,x);
coma=sqrt(8)*(3*(r.^2)-2).*r.*cos(theta);
ap=zeros(90);
ap(36:55,36:55)=circ(10).*exp(i*2*pi*(.2)*coma);
mesh(abs(ap))                will show a flat top
                             cylinder (Figure
                             9.24(e)).
mesh(angle(ap))              will show a coma-
                             aberrated pupil,
```

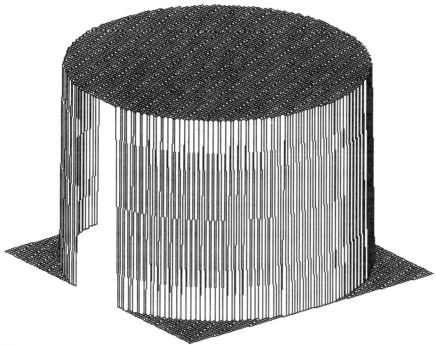

Figure 9.24(b) The round exit pupil into which we shall put the coma-generated phase aberration.

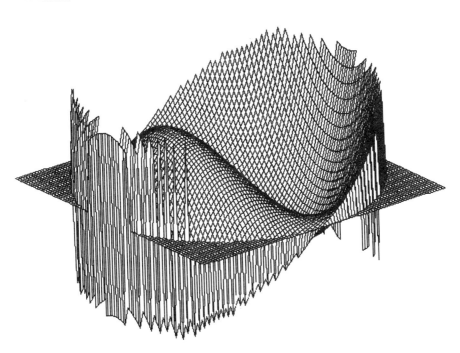

Figure 9.24(c) Representation of the coma-aberrated wave front as it would emerge from the exit pupil.

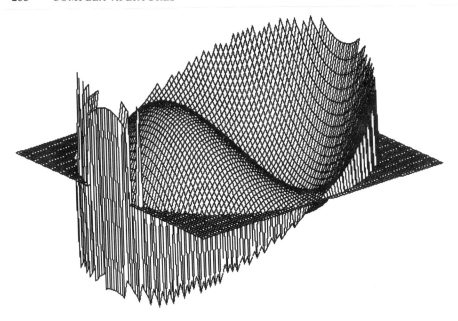

Figure 9.24(d) Figure 9.24(c) improved somewhat.

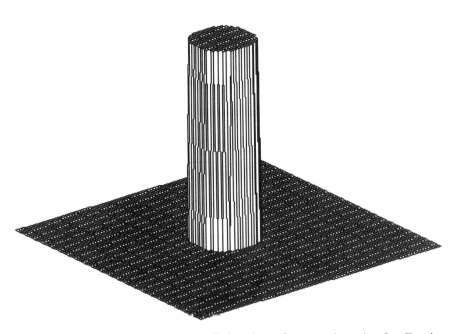

Figure 9.24(e) Creating a smaller pupil function of appropriate size for Fourier transforming in MATLAB. This is the absolute value of the phase-aberrated pupil function.

MORE COMPLEX APERTURE FUNCTIONS; ABERRATIONS   267

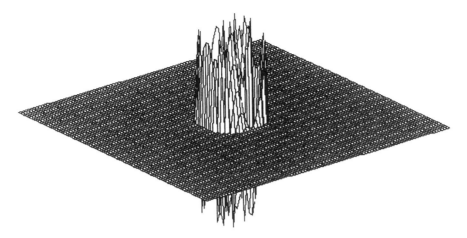

Figure 9.24(f) The coma phase variation in the smaller pupil. At this scale the pupil function doesn't look too good.

Figure 9.24(f), but perhaps not well shown at this size, but of appropriate size for Fourier transforming and squaring for the *PSF*. Transferred to SURFER and played with there for tilt, rotation, use of *X*, *Y*, and *Z* lines, size changes, etc., for a better view: Figure 9.25 is the pupil function for coma; Figure 9.26 is the irradiance, the *PSF*; Figure 9.27 is the modulus.

For the *OTF* we'll use a larger aperture and discrete convolution.

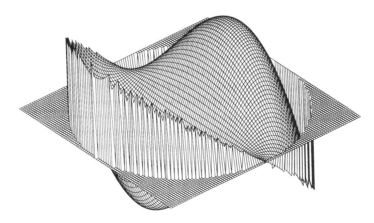

Figure 9.25 The coma pupil function transferred to and altered in SURFER.

268 COMPLEX APERTURES

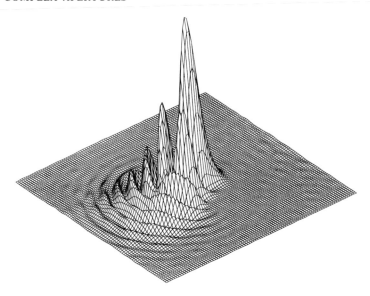

Figure 9.26 The Fraunhofer irradiance for the coma-aberrated pupil function, the point spread function, *PSF*.

```
[x,y]=meshdom(-1:2/44:1,-1:2/44:1);
r=sqrt(x.^2+y.^2);
theta=atan2(y,x);
coma=sqrt(8)*(3*(r.^2)-2).*r.*cos(theta);
ap=circ(22.5).*exp(i*2*pi*(.2)*coma);
comaotf=conv2(ap,ap);          whos tells us it is
                               complex.
comamtf=abs(comaotf);          creates the modulus,
                               the modulation
                               transfer function,
                               MTF.
comaptf=angle(comaotf);
```

this last creates *PTF*, the phase transfer function. Having done the convolution of the pupil function with MATLAB, we also processed it in SURFER. Figure 9.28 is the *MTF* and Figure 9.29 is the phase transfer function (*PTF*), which we smoothed in SURFER. How's that!

All of these figures, especially the phase plots, require some thoughtful interpretation.

9.2.2 Astigmatism

We can examine the astigmatism aberration in the same manner this time using the Zernicke Polynomial Number 5 from the table of Kim and Shannon.

MORE COMPLEX APERTURE FUNCTIONS; ABERRATIONS 269

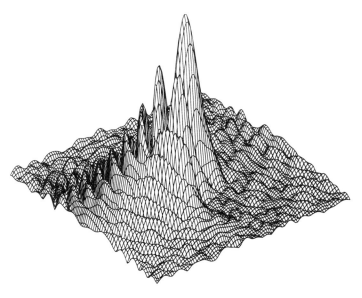

Figure 9.27 The modulus of the Fourier transform of the coma-aberrated pupil function.

We have somewhat changed the view tilt and rotation from that used by Kim and Shannon.

Figure 9.30 is the deviation from spherical of the aberrated wave front as it emerges from the exit pupil. The modulus of the complex Fourier transform is Figure 9.31; you can see that there is some light all over the place. The

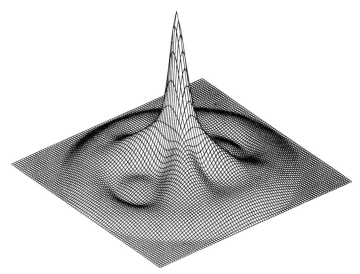

Figure 9.28 The modulation transfer function, MTF, for the coma-aberrated pupil function.

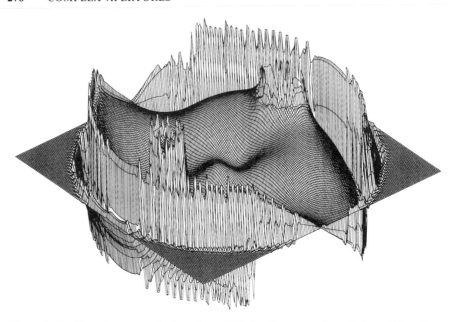

Figure 9.29 The phase transfer function, *PTF*, for the coma-aberrated pupil function.

irradiance in the Fourier transform plane is displayed as Figure 9.32; compared to the central region, there is not much light away from the optical axis.

The convolution of the aperture function (the exit pupil wave-front deviation from spherical function) yields a complex *OTF*; the modulus, the *MTF*, is Figure 9.33, and the phase transfer function is shown in Figure 9.34; again the steps are π.

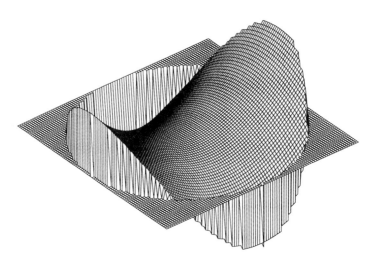

Figure 9.30 Aberrated by astigmatism, this is the departure from a perfectly spherical wave front on emergence from the exit pupil

MORE COMPLEX APERTURE FUNCTIONS; ABERRATIONS 271

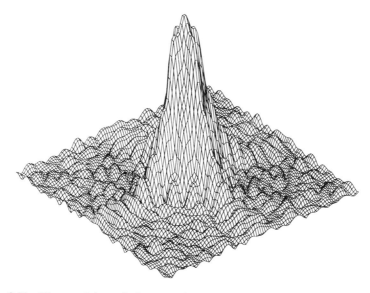

Figure 9.31 The modulus of the complex Fourier transform of the astigmatism-aberrated pupil function.

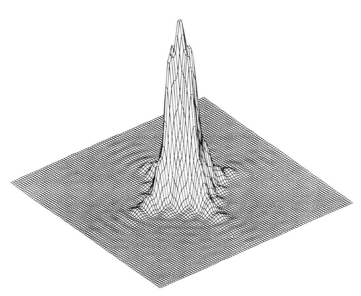

Figure 9.32 The Fraunhofer irradiance for the astigmatism-aberrated pupil function, the point spread function, *PSF*.

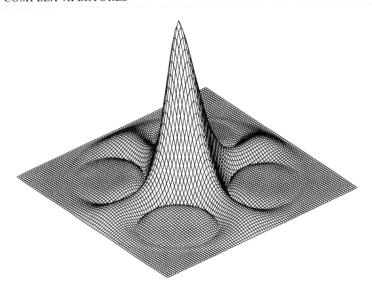

Figure 9.33 The modulation transfer function, *MTF*, for the astigmatism-aberrated pupil function.

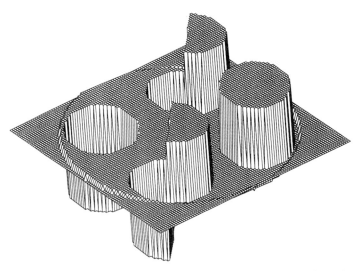

Figure 9.34 The phase transfer function, *PTF*, for the astigmatism-aberrated pupil function.

9.2.3 Focus Error

In a similar manner, using the focus-error polynomial, Number 4 of Kim and Shannon, Figure 9.35 is the deviation from spherical of the aberrated wave front upon emergence from the exit pupil—what we have usually called the aperture function. Figures 9.36 through 9.39 show: the

MORE COMPLEX APERTURE FUNCTIONS; ABERRATIONS 273

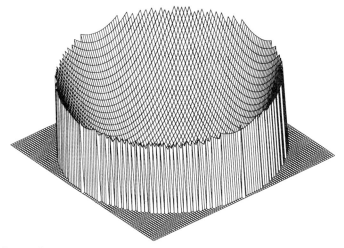

Figure 9.35 Deviation from spherical of the aberrated wave front, for focus error, upon emergence from the exit pupil.

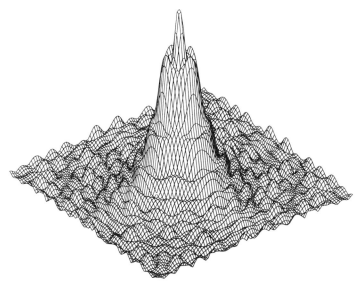

Figure 9.36 The modulus of the complex Fourier transform of the pupil function for focus error.

modulus of the Fourier transform, the irradiance (the *PSF*), the *MTF*, and the *PTF*.

9.2.4 Zernicke Polynomial Number 34

Zernicke Polynomial Number 34 is also a combination of different amounts of various orders (powers) of the primary aberrations. The deviation from

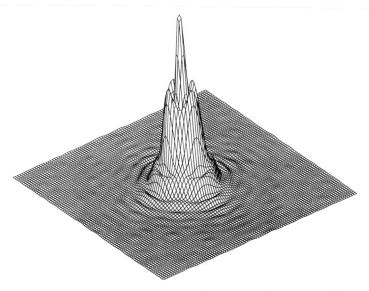

Figure 9.37 The Fraunhofer irradiance for the focus-error-aberrated pupil function, the point spread function, *PSF*.

spherical of the pupil function looks quite complex (Figure 9.40). Figures 9.41 through 9.44 show: the Fourier transform modulus, the *PSF*, the *MTF*, and the *PTF*. (Wish I had been able to display these when I first began teaching optics.)

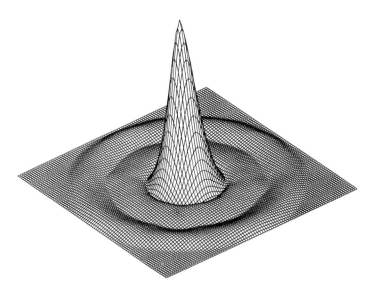

Figure 9.38 The modulation transfer function, *MTF*, for the focus-error-aberrated pupil function.

MORE COMPLEX APERTURE FUNCTIONS; ABERRATIONS   275

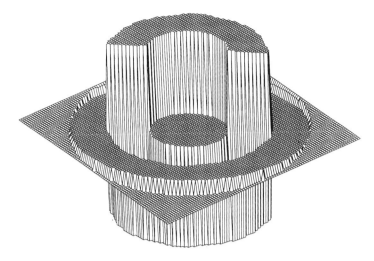

Figure 9.39 The phase transfer function, *PTF*, for the focus-error-aberrated pupil function.

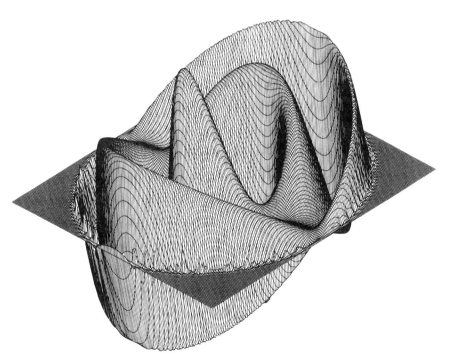

Figure 9.40 The phase error in the combination of aberrations referred to as Zernicke Polynomial Number 34.

Figure 9.41 The modulus of the complex Fourier transform of the pupil function for Zernicke Polynomial Number 34.

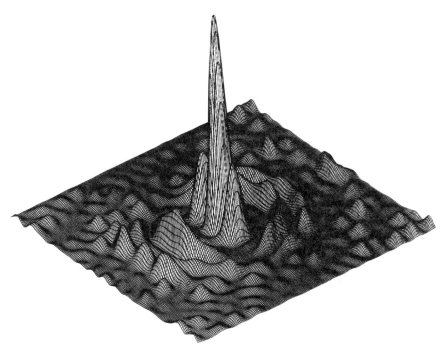

Figure 9.42 The Fraunhofer irradiance for the Zernicke Polynomial Number 34 aberrated pupil function, the point spread function, *PSF*.

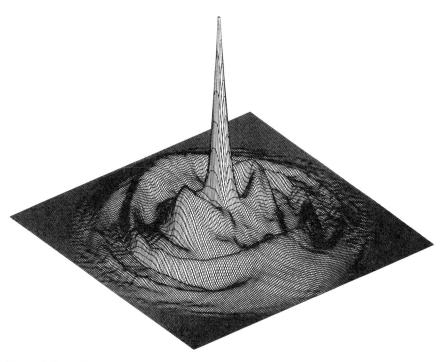

Figure 9.43 The modulation transfer function, *MTF*, for the Zernicke Polynomial Number 34 aberrated pupil function.

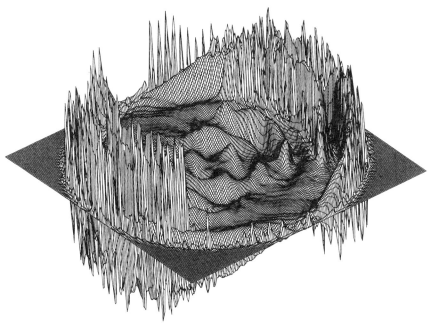

Figure 9.44 The phase transfer function, *PTF*, for the Zernicke Polynomial Number 34 aberrated pupil function.

278 COMPLEX APERTURES

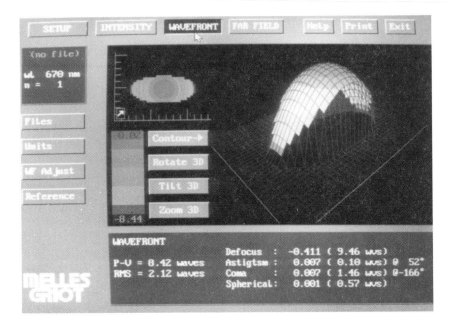

Figure 9.45 The Melles Griot *WaveAlyzer*™ computer and software system can display an optical wave front (the phase surface). (Photo courtesy of the Melles Griot Company of Boulder Colorado, and offices throughout the world.)

Interferometers can create images in the laboratory which "show" the phase, or phase error, in a wave front. The Melles Griot Company has recently created an optical instrument linked with a computer and software which can map out in "meshlike" fashion, as we have been doing, wave-front phase as well as intensity. Figure 9.45 is a screen image of the Melles Griot *WaveAlyzer*™ mapping and analyzing the amounts of specific aberrations in this particular wave front.

9.3 A MORE COMPLETE SUMMARY OF BASIC FOURIER THEOREMS

$$\text{If } A(x, y) \xrightarrow{\text{Fourier transforms to}} E(\xi, \eta)$$

$$\text{and } B(x, y) \xrightarrow{\text{Fourier transforms to}} F(\xi, \eta)$$

9.3(a) Linearity Theorem

$$A(x, y) + B(x, y) \xrightarrow{\text{Fourier transforms to}} E(\xi, \eta) + F(\xi, \eta).$$

A MORE COMPLETE SUMMARY OF BASIC FOURIER THEOREMS   279

To make this more complete, it is then also true that for all *complex* numbers p and q,

$$pA(x, y) + qB(x, y) \xrightarrow{\text{Fourier transforms to}} pE(\xi, \eta) + qF(\xi, \eta)$$

9.3(b) Shift Theorem

For real numbers s and p

$$A(x \pm s, y \pm p) \xrightarrow{\text{Fourier transforms to}} e^{\pm i 2\pi(s\xi + p\eta)} E(\xi, \eta)$$

9.3(c) Modulation Theorem

For *real* numbers s and p

$$A(x, y) e^{+i 2\pi(s\xi + p\eta)} \xrightarrow{\text{Fourier transforms to}} E(\xi - s, \eta - p)$$

9.3(d) Scaling Theorem

For *real* numbers $s > 0$ and $p > 0$

$$A(sx, py) \xrightarrow{\text{Fourier transforms to}} \frac{1}{sp} E\left(\frac{\xi}{s}, \frac{\eta}{p}\right)$$

9.3(e) Separation Theorem

If $A(x, y)$ can be written as $f(x)g(y)$, and if $f(x)$ Fourier transforms to $F(\xi)$ and $g(y)$ Fourier transforms to $G(\eta)$, then

$$A(x, y) \xrightarrow{\text{Fourier transforms to}} F(\xi)G(\eta)$$

(Note: $f(x)$ is a function of x only, $g(y)$ a function of y only.)

9.3(f) Radial Functions

With $r = \sqrt{x^2 + y^2}$, if $A(x, y)$ can be written as $A(r)$ and $A(r)$ is Fourier transformable, then

$$E(\rho) = 2\pi \int_0^{+\infty} A(r) J_0(2\pi \rho r) r \, dr$$

where $\rho = \sqrt{\xi^2 + \eta^2}$.

COMPLEX APERTURES

These theorems and relationships are derived or proven in many books on Fourier analysis but sometimes for functions only of one-dimension. We have found Walker (1988), Gaskill (1978) and Goodman (1968) useful.

REFERENCES FOR WAVE ABERRATIONS

Ditchburn (1976); Hopkins (1950); Kim and Shannon (1987); Mahajan (1991); Nussbaum and Phillips (1976); O'Neill (1963); Smith and Thompson (1971); Welford (1986); Williams and Becklund (1972), (1989).

FOR FURTHER CONSIDERATION

1. In Section 9.1.3 we illustrated the diffraction process if the light incident on the aperture is still a plane wave, but not perpendicular to the aperture plane, i.e., the phase varied linearly across the aperture. Try to describe the situation, calculate the Fourier transform, if other, nonplanar waves, are incident on the aperture plane, say cylindrical or spherical wave fronts.

10

OPERATIONS IN THE FOURIER TRANSFORM PLANE

In a very early chapter we illustrated the filtering of a square wave in 2-space; in several examples we selectively removed high frequency constituents of a square wave and in a last example we removed only the low frequency fundamental. When these sorts of operations are carried out on electrical signals, say in a radio or audio amplifier circuit, they are referred to respectively as low-pass filtering and high-pass filtering; in the first instance passing the low frequencies on to the output stage while channeling the higher frequencies to electrical earth ground. High-pass filtering is the inverse, allowing only high frequencies to be passed on to the output stage. A third type of filter could be created, a band-pass filter, whereby both low and high frequencies are channeled to earth ground while a band of middle frequencies go to output. There are similar operations with light.

For visible light a red camera filter is a low-pass filter allowing low frequency, long wavelength red light to pass through while, in this case, absorbing the higher frequencies. A violet or perhaps blue filter would be considered a high-pass filter, while a yellow–green filter might be considered a band-pass filter. Over a larger spectrum which included the ultraviolet and infrared frequencies, all these filters would be band-pass filters.

Filtering can be conceived for all kinds of waves; sound waves in air come to mind—the distant sound of thunder; listen very carefully sometime. Shoreline structures could perhaps be designed to allow in those water waves which leave sand rather than remove it, hence building beaches rather than destroying them. Gravitational waves, seismic waves, periodic fluctuations in climate (though not a wave), tsunamis, tidal waves, etc., contain properties which could be analyzed for "harmonic" content. Should we include "waves of violence" which erupt in nations and propagate from one city and perhaps to another; the list could go on. You see, filtering also occurs outside the laboratory. Can you conceive of filtering techniques for each?

In a less technical sense, communications in general can be considered wave and information propagation. Think of the many "filtering processes" that occur when messages are sent, e.g., the children's game "telephone" or "Pass it on;" or perhaps this example from Mosley's 1966 biography of Hirohito, filtering through a "decoding process," of a message which one might hope was not aberrated at its beginning:

> Not for the first time, however, a high Japanese official failed to translate a plain American offer into plain Japanese. By the time (Ambassador to the United States—Admiral) Nomura had rendered the President's (Roosevelt's) proposal into his own language, it no longer made any sense and neither (Foreign Minister) Toyoda nor (Prince Konoye—Prime Minister) Konoye could make head or tail of it. Not until July 27 (1941) when (American Ambassador to Japan) Grew saw the Japanese Foreign Minister in Tokyo and repeated the President's offer did its meaning become clear. But by then it was too late. The Japanese government had already officially announced its decision to occupy Indochina (formerly French Indochina)—and it would have meant a loss of face to abandon its plans. The deed was done. It was to have momentous consequences. (added parentheses)

Yes, signal filtering occurs outside the lab.

Our Fourier optical transforms contain, "in code," information about the optical system from which they have emerged. We have learned that an inverse transformation will yield back some of that information, but the phase information usually presents quite a battle. Indeed, when x-rays emerge from a system containing the three-dimensional structure of important molecules, the decoding can be quite difficult especially for molecules with molecular weights in excess of, or some multiples of, 10,000. Luckily for most of us there is not a potential World War dependent upon our success at correct decoding of messages in two-dimensional light beams. But isn't a returned radar signal very similar to our Fraunhofer diffraction problems except that the aperture, an aircraft, now represents a reflection aperture. It seems that all the elements of amplitude transmission and phase are present, and it seems also that this is starting to look like holography, a subject we leave to other books. Decoding messages in diffracted light beams will be very important in medicine, for discovering gene and virus structures.

In more everyday occurrences even the audio and visual images of one's self are filtered so that we do not get correct images. Our own voices sound different to us than to others because to us the sound is filtered through bone, muscle and resonant structures in the head. Just listen to yourself sometime, played back from a well-made audio tape. Each morning we see ourselves not as others see us but our image is filtered by a strange device which performs a left/right reversal; I don't part my hair on the right though my mirror says I do. But I have become content with that "backward" image of myself. The image we have of ourselves as teachers, filtered through our own egos, is often very different from the image recorded on videotape from the back of the room.

In some of the above examples it is not immediately clear what form should be used to describe the constituent parts of the "messages" or "information." Is there a "spectrum" in each case, and if so, what is its nature? How do filtering processes take place? And it is clear that filtering can degrade a message—filtering removes information; but we shall see that it can also enhance a message—e.g. filtering can remove undesirable information.

10.1 STEP INTO THE FOURIER TRANSFORM PLANE

Now, in this book, we find ourselves with access to the spatial frequency spectra and phase spectra of two-dimensional spatial apertures and two-dimensional spatial and phase object scenes; spectra which can be displayed or at least be present in the rear focal plane of a lens. What an excellent opportunity! These are some of the things we can do to the spectra: block out some spatial frequencies, highs, lows, both highs and lows, or a band; block out only those in, say, the "+" direction; with a transmission filter alter the amplitude of selected frequencies; with a phase filter alter the phase of selected frequencies; block out all but selected frequencies which we know correspond to a particular signal for which we might be searching; create filters which can be changed in real-time to filter object scenes which are changing in real-time; by trial and error, by guess and by golly, block out and alter amplitudes and phases to "enhance" the information being sought in an image; the list can go on. If white light is diffracted by the object scene, certain spatial frequencies can be "colored" by placing different color filters in different parts of the Fourier transform plane.

The spatial frequency spectrum of the object just sits there in the rear focal plane of the "transform" lens, floating in space, carrying with it the subtle nuances of phase, and called by many "the Fraunhofer diffraction pattern of the object." Periodic spatial structures in the object give rise to regular patterns in the diffraction pattern. We have come to understand that light from high spatial frequency information in the object scene, corresponding to fine detail in the object, is diffracted out at relatively large angles to the optical axis and is present in the Fourier transform plane displaced considerably from the optical axis. Correspondingly, larger structures in the object scene of low spatial frequency, diffract light at smaller angles which then pass on through the transform plane close to the optical axis. In fact, the largest spatial structure in our diffracting apertures (objects) are responsible for the smallest spots of light in the diffraction pattern. Remember that inverse relationship: small objects diffract light at large angles and vice-versa.

10.2 EDGE DETECTION BY SPATIAL FILTERING

An example of filtering of spatial frequencies is given in Figure 10.1 which represents a square aperture blocked by an opaque triangle. We will attempt to do edge detection. This aperture is placed on the optical axis in the front focal plane of a lens and illuminated with coherent light, the plane waves of a

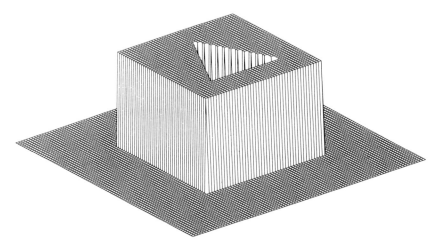

Figure 10.1 Square aperture blocked by an opaque triangle.

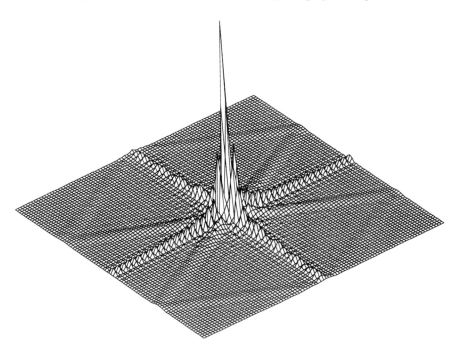

Figure 10.2 Modulus of the Fourier transform of Figure 10.1.

laser beam. The information about the tiny detail of the sharp edges is in the high spatial frequencies of the Fourier transform, i.e., diffracted at large angles. The modulus of the transform of this aperture is shown in Figure 10.2. We now go into the computer-generated transform matrix in MATLAB and block out (set to zero) a square array of all the low spatial frequency terms, see Figure 10.3 (which again shows the modulus) in which there has been a vertical scale change. This corresponds to putting a small opaque square on the optical axis in the lens rear focal plane, the Fourier transform plane. Now Fourier transform the transform matrix, which is equivalent to reimaging the aperture with another lens; the modulus is Figure 10.4. What one would see or detect (the

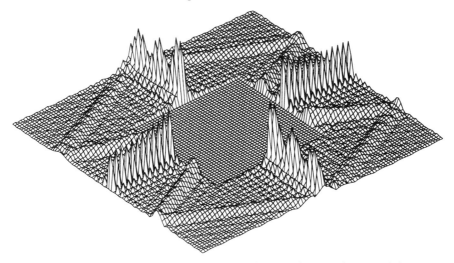

Figure 10.3 Blocking low spatial frequencies, Fourier transform modulus.

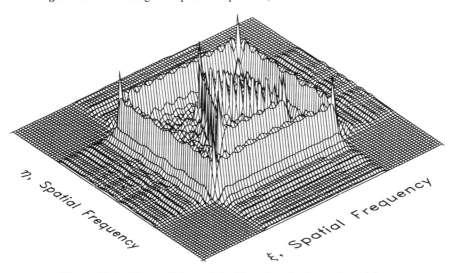

Figure 10.4 Figure 10.1 spatially filtered and reimaged modulus.

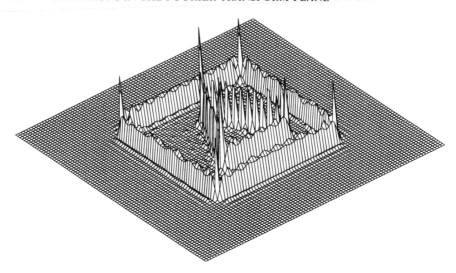

Figure 10.5 Figure 10.4 irradiance; edge detection of Figure 10.1.

reimaged but now filtered aperture) is the complex square of this transform, see Figure 10.5, wherein we would see brightness mainly along the edges of the square aperture and its blocking triangle. We think this is rather neat! In SURFER we can stack the basics of this process (Figure 10.6). Have you ever tried demonstrating spatial filtering to a large group of people? These images can be used as slides or overhead transparencies.

10.3 SPATIAL FILTERING FOR DETECTION OF APERTURE PHASE

In Section 9.2.4 we examined the phase aberration given by Zernicke Polynomial Number 34. Let's see what would happen if we spatially filtered the Fourier transform of this clear but phase-rich aperture. Figure 10.7 shows the phase of the aperture. If we image this uniformly illuminated aperture, the irradiance looks like Figure 10.8, just a "clear" aperture. Remember that the phase information is lost in the irradiance. Though we can't see it, Figure 10.9 is the modulus of the Fourier transform and it is "present" in the rear focal plane of our transform lens. Figure 10.10 is the irradiance in the transform plane and you can see that we have reached in and blocked half of the transform from getting through to the reimage of the aperture. Sounds like a Foucault knife edge doesn't it? The filtered reimaged aperture has the irradiance of Figure 10.11, where we now see some bright and dark structure, in contrast to Figure 10.8. But it doesn't look like the phase pattern of Zernicke Polynomial Number 34; and it should not. Figure 10.7 was phase *amplitude*! Figure 10.11 is amplitude squared. So let's see what Figure 10.7 looks like, squared. There it is, Figure 10.12.

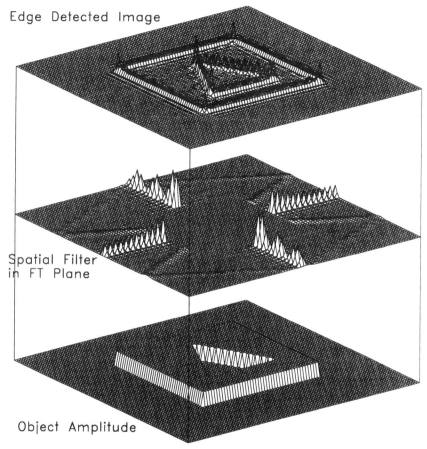

Figure 10.6 SURFER stacked essentials of the process demonstrated in Figures 10.1 to 10.5.

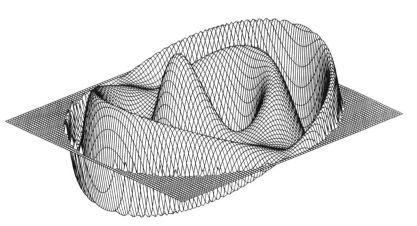

Figure 10.7 Aperture phase represented by Zernicke Polynomial Number 34.

288 OPERATIONS IN THE FOURIER TRANSFORM PLANE

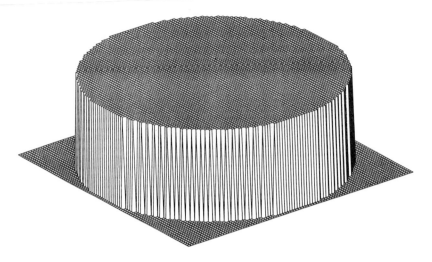

Figure 10.8 Irradiance image of the aperture of Figure 10.7.

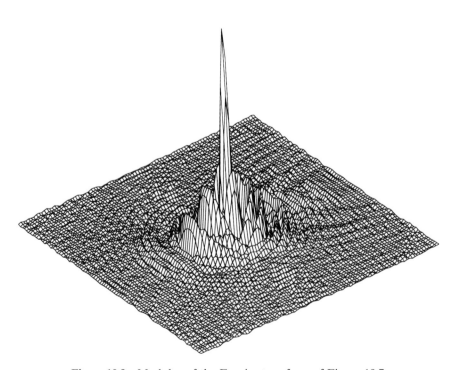

Figure 10.9 Modulus of the Fourier transform of Figure 10.7.

SPATIAL FILTERING FOR DETECTION OF APERTURE PHASE 289

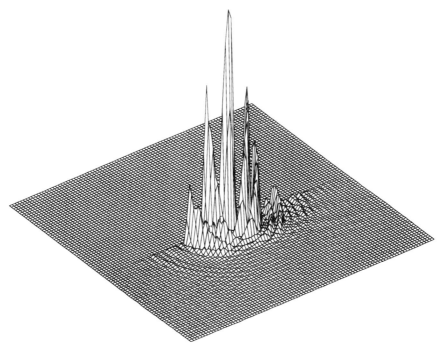

Figure 10.10 Irradiance in the Fourier transform plane of Figure 10.7, showing half of the transform blocked.

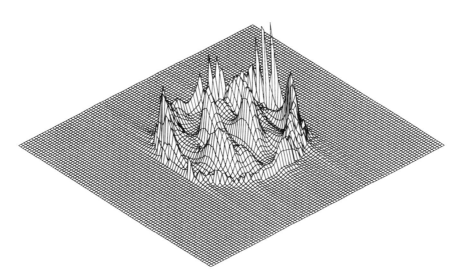

Figure 10.11 Irradiance in the filtered reimage of the aperture of Figure 10.7.

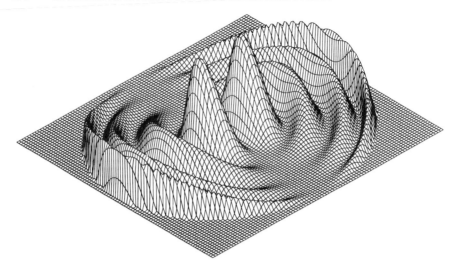

Figure 10.12 The complex square of the aperture of Figure 10.7.

Figure 10.11 does not have the structure of the Number 34 polynomial, Figure 10.12. That is because we carried out nonsymmetrical filtering. We did get evidence of some sort of phase structure in the aperture, but we took information from only one side of the transform. A similar procedure does occur in Foucault knife-edge testing. In order to get proper information about the phase spacing, size, and distribution, we require information from opposite sides of the transform. What we should block out, symmetrically, is the central maximum, the low spatial frequency information, and image with just the high spatial frequency details of the aperture shape and phase.

10.4 RESULTS FROM THE SPATIAL FILTERING LABORATORY

Although we have done some spatial filtering in our labs, this author can by no means claim that his results, for the purposes of this book, are anywhere near as satisfactory as the excellent results obtained by: H. Lipson and C. A. Taylor (1958); C. A. Taylor and H. Lipson (1964); H. Lipson and colleagues (1972); and G. Harburn, C. A. Taylor, and T. R. Welberry (1975); and their colleagues; work which must span well over 20 years. Plates 30 and 31 from the *Atlas of Optical Transforms* by G. Harburn, C. A. Taylor, and T. R. Welberry, originally published by G. Bell & Sons Ltd in London, and now handled by Routledge in London, are reproduced here as Figures 10.13(a) and (b) and 10.14(a) and (b). We added parenthetical expressions to the figure captions from the *Atlas*. The reader is advised to examine these Figures now, then read on for additional interpretive information.

Central dark-ground microscopy, referred to in the caption for Figures 10.14(a) and (b), places a small opaque obstruction in the transform plane to block out only the central portion of the transform. Thus all the detail, the high spatial frequencies of the object, are used to form the image whereas the low, and "dc" spatial frequencies, the "background illumination," is eliminated. So for that piece of mica (a phase object), all the phase boundaries (phase steps) show up in the filtered image. And, of course, only high spatial frequencies go into making up the last image of the mouse.

Notice in image 7 of Figure 10.14(a) that only information in the η-direction was allowed through the transform plane, hence spatial detail in only the Y-direction was recovered in the image.

In Figure 10.13, to get the bar structures properly imaged, the related spatial frequencies must make it through the transform plane. In image 4(a), the central region of $E(\eta, \xi)$ was blocked out, leaving a larger separation of information about spatial detail in the η-direction. This was interpreted in the $X-Y$ plane as corresponding to larger spatial frequencies in the object; the image bars are twice as frequent. Such comments may help you interpret the remaining sets of figures.

The other plates in the *Atlas* are equally informative, some contain phase examples; for some you might wish to attempt computer simulation. The *Atlas* is the only book I have ever read in which the Appendices immediately followed the Introduction. Figure 10.14(c) is the well thought out optical system they employed for these pictures.

10.5 AN INVERSE PROBLEM

Some detective work can be done. Much to the embarrassment of the author an error was made in two graphics used in another publication, in the early printings. It was supposedly the Fraunhofer amplitude of a cross. It was a simple job for SURFER: sum my calculated Fourier transforms for two perpendicular 3×1 slits, subtract the transform for one centered 1×1 square, and simplify. We misplaced a factor 3 and came up with an incorrect transform and diffraction pattern. We spotted the error comparing figures after MATLAB did the transform of the cross.

But the error raised the question "To what aperture did the incorrect transform belong?" The erroneous calculation, purely real, was put into the MATLAB matrix (easy to do). Inverse transformation should yield the "mystery" aperture. So the inverse Fourier transform **ifft2** was performed which gave a complex aperture. Figure 10.15 is the modulus, which has some phase noise. The phase pattern was even more difficult to interpret. This is an example of an "inverse" problem, made easier because we had the Fourier transform and not its square, the irradiance. Phase information is lost in the squaring process. When one looks at the x-ray diffraction pattern of, say, a biologically important molecule, the question is inverse also, "What structure caused that?" When the radar

292 OPERATIONS IN THE FOURIER TRANSFORM PLANE

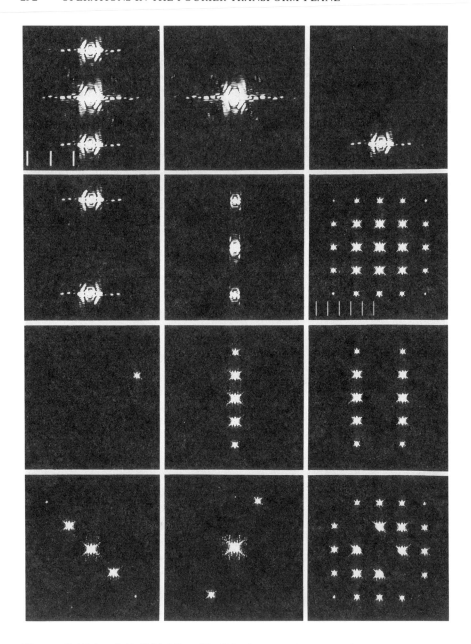

Figures 10.13(a) and 10.13(b) (Plate 30 of Harburn *et al* (1975) "The left-hand page (a) shows examples of diffraction patterns after modification by various obstructions (spatial frequency filters). The right-hand page (b) shows the images formed ... from the information remaining in the diffraction patterns (images formed by a lens suitably placed after the transform plane where the filtering was performed). (Counting across),

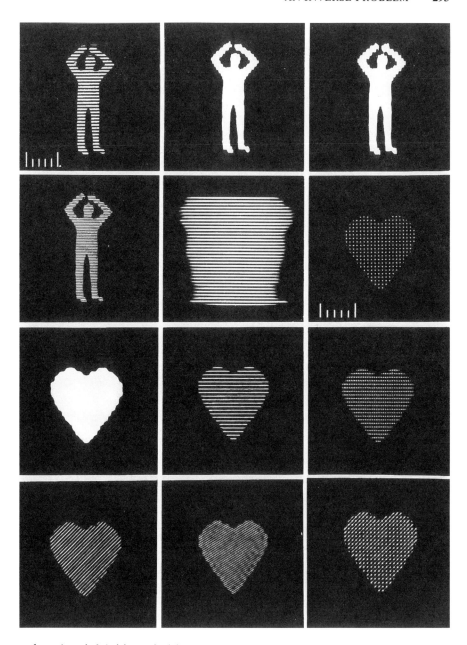

numbers 1 and 6 (with vertical bars) show the complete diffraction patterns and the corresponding 'perfect' images." We have added parenthetical expressions. (G. Harburn, C. A. Taylor, and T. R. Welberry, *Atlas of Optical Transforms*, © 1975, reproduced with permission of the publisher, Chapman and Hall in London, originally published by G. Bell & Sons in London.)

294 OPERATIONS IN THE FOURIER TRANSFORM PLANE

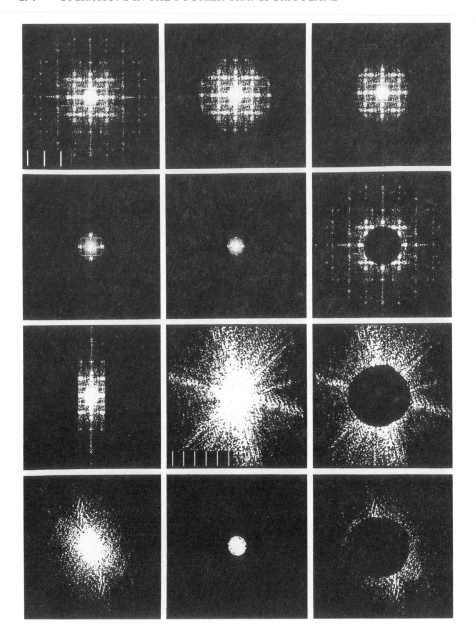

Figures 10.14(a) and 10.14(b) (Plate 31 of Harburn *et al* (1975) "(Counting across), numbers 1 and 8 (with vertical bars) and 10 (the mouse) are complete diffraction patterns. Numbers 1–7 show the effect on resolution of excluding information from the final image for a test object. The amount of error in image 4 is particularly interesting. The diffracting object for 8 was a piece of mica with steps and scratches on the surface

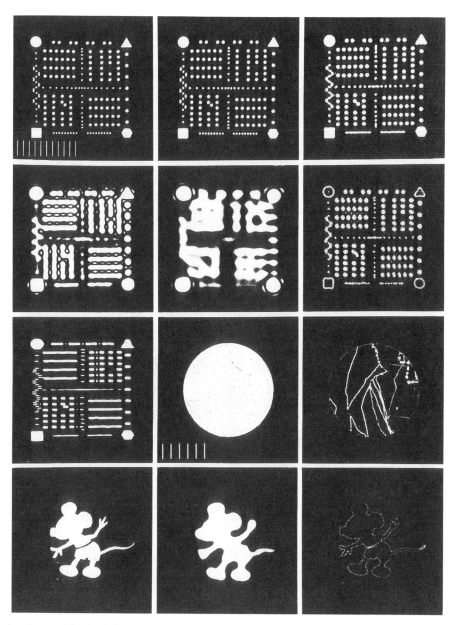

(a phase object); 9 is an example of central dark-ground microscopy." We have added parenthetical expressions. (G. Harburn, C. A. Taylor, and T. R. Welberry, *Atlas of Optical Transforms*, © 1975, reproduced with permission of the publisher, Chapman and Hall in London, originally published by G. Bell & Sons in London.)

296 OPERATIONS IN THE FOURIER TRANSFORM PLANE

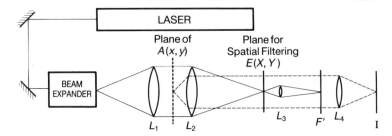

Figure 10.14(c) The laser, beam expander, and L_1 produce monochromatic plane waves on $A(x, y)$. $E(X, Y)$ is in the rear focal plane of L_2; this is the plane where the Fourier transform may be filtered with suitable "masks." L_3 is a projection lens to produce an enlarged image of $E^2(X, Y)$ on film at F'. With F' film and L_3 removed, L_2 and L_4 can form spatial frequency filtered images of $A(x, y)$ at I. (Adapted from G. Harburn, C. A. Taylor, and T. R. Welberry, *Atlas of Optical Transforms*, © 1975, G. Bell & Sons Ltd, London.)

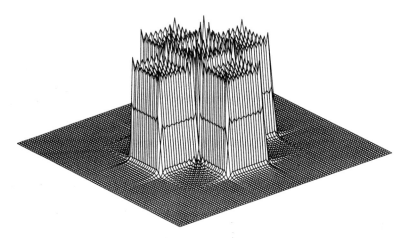

Figure 10.15 The modulus of the "mystery aperture" created in MATLAB by inverse Fourier transformation of a real but incorrectly calculated expression. See text.

signal returns, bounced off the aircraft, the question is, "What aircraft is that?" Using lasers in diffraction and interference experiments of physics courses everyone knows 632.8 nm is the wavelength. I usually ask my students to fully determine the structure causing the diffraction pattern. "Inverse problems" is a very active area of research today, and if you have access to a program like MATLAB you might want to try your hand at designing Fourier transforms with special point spread functions, and doing inverse transforms to see if the aperture is physically realizable.

Incidentally, Figure 10.16 is the cross aperture; 10.17 is its convolution; 10.18 is the transform, $E(\xi, \eta)$; 10.19 is $|E(\xi, \eta)|$; and 10.20 is the irradiance function.

A close relative of the cross aperture is this well-proportioned **T** aperture

AN INVERSE PROBLEM   297

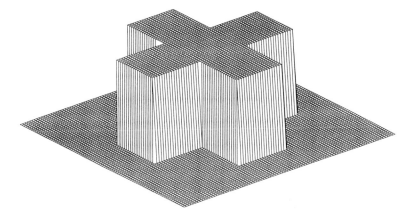

Figure 10.16 A "cross" aperture, created in MATLAB.

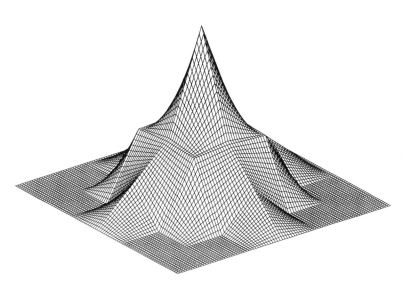

Figure 10.17 Done in MATLAB, the convolution of the "cross" aperture function of Figure 10.16 with itself.

298 OPERATIONS IN THE FOURIER TRANSFORM PLANE

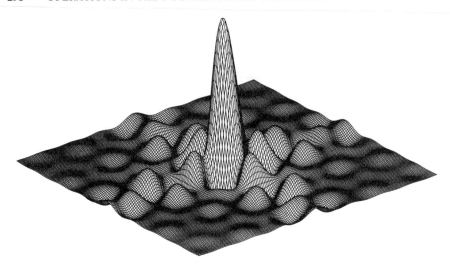

Figure 10.18 $E(\xi, \eta)$, the Fourier transform of the "cross" aperture of Figure 10.16. Can't see some of the negative values of $E(\xi, \eta)$ so go to Figure 10.19.

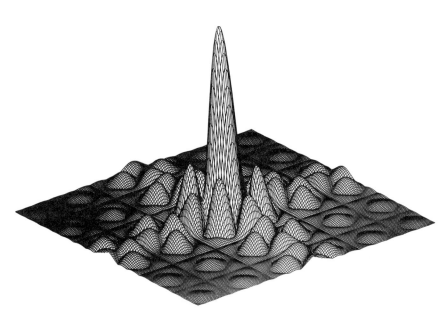

Figure 10.19 $|E(\xi, \eta)|$, the absolute value of the Fourier transform of the "cross" aperture of Figure 10.16. Now we can see the negative values, but how could one tell which, originally, were the negative values and which were the positive ones? This is called a "phase retrieval" problem, not too difficult here.

AN INVERSE PROBLEM 299

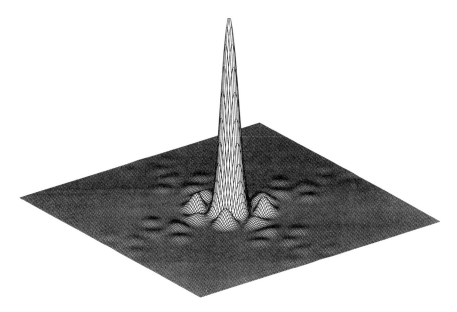

Figure 10.20 $[E(\xi, \eta)]^2$, the irradiance function in the Fraunhofer diffraction pattern for the "cross" aperture of Figure 10.16.

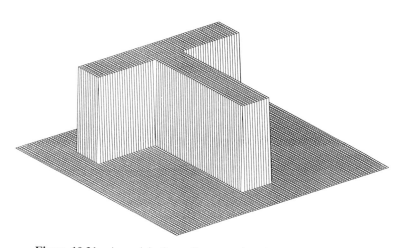

Figure 10.21 A model of a well-proportioned T-shaped aperture.

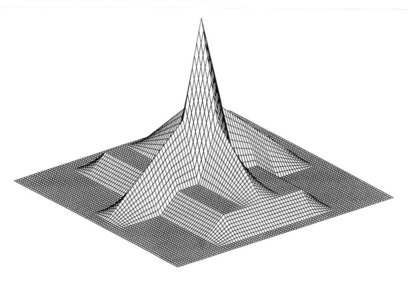

Figure 10.22 The convolution of the function of Figure 10.21 with itself.

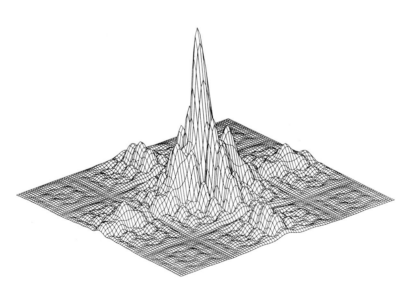

Figure 10.23 The modulus of $E(\xi, \eta)$, the Fourier transform of the function of Figure 10.21.

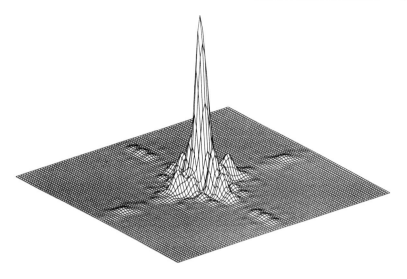

Figure 10.24 $[E(\xi, \eta)]^2$, the irradiance function in the Fraunhofer diffraction pattern for the T-shaped aperture of Figure 10.21.

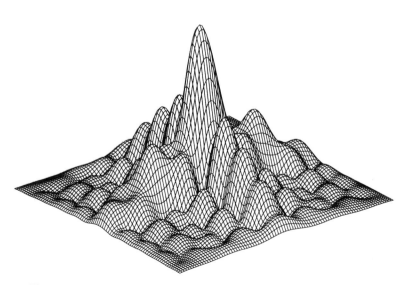

Figure 10.25 The modulus, after zoom-in and smooth, in SURFER.

302 OPERATIONS IN THE FOURIER TRANSFORM PLANE

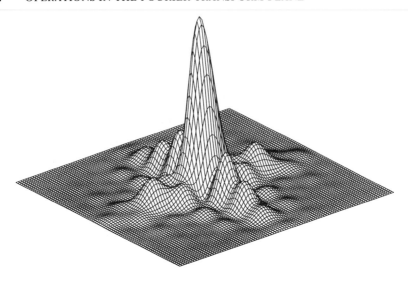

Figure 10.26 The irradiance, after zoom-in and smooth, in SURFER.

(Figure 10.21). Its convolution is 10.22; 10.23 is the modulus of $E(\xi, \eta)$; 10.24 is the irradiance function. Figure 10.25 is a "zoom-in" and "smooth" of the modulus; while 10.26 is a "zoom" and "smooth" of the irradiance.

10.6 SOME PERSONAL CONCLUSIONS

In physics instruction the somewhat traditional phasor approach to diffraction and slit interference is quite limited for continued or later usefulness. *In a general physics course where students know the rudiments of calculus, Fourier transform techniques should be an attractive alternative because a Fourier approach will have widespread usefulness for students in many fields of science and engineering.* Reasonably simple apertures Fourier transform with some work but with relative ease. We believe we have done this successfully when teaching physics to biology students; we had biological structures in mind.

We hope we have shown that simulations and visualizations of optical transformations in 3-space for any aperture can now be achieved easily on a PC. In an advanced laboratory it would not be too difficult for students to design their own diffraction apertures, for photographic reduction on, say, Kodak Ultratec™ film; and with laser and optical bench create the *PSF* for comparison with a computer simulated *PSF*. Even phase steps can be included; see the paper of H. Stark, W. R. Bennett, and M. Arm (1969). Kodak Technical Pan™ film and a range of exposures might help visualize the detail in the *PSF*'s. See the paper of S. A. Dodds (1990) in the references. For interesting aperture possibilities see: Gillon (1969); Horemis (1970); Larcher (1979); Ōuchi (1977).

Specimen grids (or gratings) used in electron microscopy make fairly good diffracting apertures. They are not expensive. Consult catalogs of (all in the USA):

1. Ernest F. Fullam, Inc., 900 Albany Shaker Road, Latham, NY 12112-1491.
2. Ladd Research Industries, Inc., P.O. Box 1005, Burlington, VT 05402.
3. Ted Pella, Inc., P.O. Box 510, Tustin, CA 92681.

For screens and meshes:

4. Industrial Reproductions Inc., 100 Northeastern Blvd., Nashua, NH 03060.
5. The Gilson Co., Inc., P.O. Box 677, Worthington, OH 43085-0677.
6. ByChrome Company, Box 1077, Columbus, OH 43216.

In this book we have only scratched the surface, so to speak, of computer simulation of the wide ranging techniques of optical processing. There is much fascination to be found in aperture design, matched- and complex-filtering, optical correlating, optical computing, etc. Computer graphics allows us to visualize more than the eye can see. The interested reader might wish to consult the following references: J. C. Brown (1971); J. W. Goodman (1977); A. Hairie and J. Provost (1983); J. L. Horner and Peter D. Gianino (1984); D. McLachlan, Jr. (1962); A. B. Meinel, M. P. Meinel and N. J. Woolf (1983); J. P. Mills and B. J. Thompson (1986); F. T. S. Yu and E. Y. Wang (1973).

11

OTHER INTERESTING AND RELATED TOPICS

11.1 LASER BEACON ADAPTIVE OPTICS

One might wish to consider an optical system some 100 kilometers long. Such is the case in astronomy where the telescope is only the last element at the bottom end of a 100 km thick multilayer stack of turbulent atmosphere. You can have the best optics in the world (referred to as diffraction limited optics; aberrations all corrected), but if the sky, the atmosphere, won't settle down and cooperate, the "seeing," the images, will be quite poor.

We know there are some things we can do to improve, or at least change, the diffraction pattern of a star. We can design for the telescope an apodizing aperture or one for super-resolution (see Chapter 8). But how could we change multiple atmospheric layers, 100 km of them, bouncing around all night, bending wave fronts from a star this way and that? How about putting the telescope at the bottom end of a 100 km long glass top tube and pump all the air out? Hmm!? Then, more practical, put the telescope up above the turbulent atmosphere. The Hubble Space Telescope is in orbit about the Earth taking pictures with its CCD detector, and from its photos you can see the evidence of digital imagery from the many pixels in the CCD. Those digitized images are a great form for telemetering back to Earth, with probably very little error. And were we not lucky to know what optical errors existed in the first Hubble, so that after receiving the faulty images, image processing by computer could remove a great deal of the error. The Hubble pictures that are published in books, journals, and magazines are great!

But suppose we could adapt our optics on the ground to correct the imagery for the distortion caused by a turbulent or disturbing atmosphere. Suppose we could hang an artificial star or test target in orbit about the Earth just outside the atmosphere. We would then know what the image was supposed to look like and we could "adapt" the optics to make it look so. Problem: the atmosphere over Mount Palomar is not the same as that over Kitt Peak or Pic du Midi, and our artificial star is in orbit, probably moving with respect to any observatory. We could put it in a stationary position about the Earth, in a Clark Orbit, just like the communications satellites which appear to stay in one position, directly over the Earth's equator. Their orbital period is 24 hours, just as the Earth's rotational period is 24 hours. They appear "parked" somewhere along a line in the line which we call the celestial equator. So that even if we could provide such an artificial star for every observatory in the world, packed on the celestial equator, they could correct their star images if they were observing stars in close proximity to the artificial ones, stuck on the celestial equator. What happens if they wish to observe objects in the other 99.99999% of their sky? No artificial star anywhere nearby.

All things considered, it seems the only choice is to put the artificial star wherever the telescope is pointing, hence the Laser Beacon concept. The laser is mounted with the telescope and points in the same direction as the target object; in fact the laser light shares the telescope optics, going out and returning. A laser beam with sufficient intensity causes the atmosphere to glow, i.e., imitate a star. At altitudes of 10–20 km, the stratosphere, the glow is caused by Rayleigh scattering, but at altitudes of 80–100 km the glow is by resonant fluorescence scattering from atomic sodium in the mesosphere. The backscattered light from such artificial stars is received by the telescope and its accompanying opto-electronics system which senses the corrections necessary, and then sends appropriate signals to the deformable "flexible" telescope mirror which adjusts its shape to produce a good image. Laser beacons can be used to correct higher-order distortions caused by the atmosphere, but the correction for full-aperture tilt of incoming wave fronts can be corrected by adaptive optics using natural stars. As you can imagine, this creates a quite complicated system, both to build and to analyze, see Figure 11.1. Figure 11.2 shows the comparison of star images taken with the *Phillips Laboratory's Starfire Optical Range* 1.5 meter telescope at Kirtland AFB, New Mexico. Furnished to us by the Technical Director, Robert Q. Fugate, we find them quite remarkable; the figure caption tells the entire story. The full width half maximum image sizes (these images are what we have called *PSF*'s) are 1.65 and 0.18 arc seconds. The laboratory uses a copper vapor laser operating at 5,000 pulses per second. For more information and more references consult the following article from which the above was obtained: R. Q. Fugate, "Laser Beacon Adaptive Optics," *Optics and Photonics News*, Vol. 4, June 1993, pp. 14–19, and the cover illustration.

Early amateur astronomers and their careful observations were able to make significant contributions to professional astronomy. But as professional telescopes became larger, and instrumentation more sophisticated, the work of amateurs was less needed. With the advent of CCD imaging equipment,

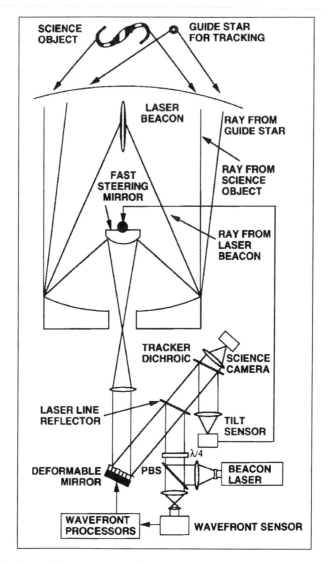

Figure 11.1 Essential components of a laser beacon adaptive optics system. In this configuration, the laser beacon shares the full aperture of the telescope by means of a narrow-band laser line filter. A polarizing beam splitter and quarter-wave plate form a duplexer, allowing the backscattered light to pass through the beam splitter to the wave front sensor. A separate tilt sensor tracks a natural guide star near the science object using the full aperture of the telescope. The secondary mirror of the telescope serves as the fast steering mirror to correct full aperture tilt. (Work of *Starfire Optical Range, Phillips Laboratory*, Technical Director: Robert Q. Fugate. Reproduced from *Optics & Photonics News*, June 1993, p. 16; © Optical Society of America.)

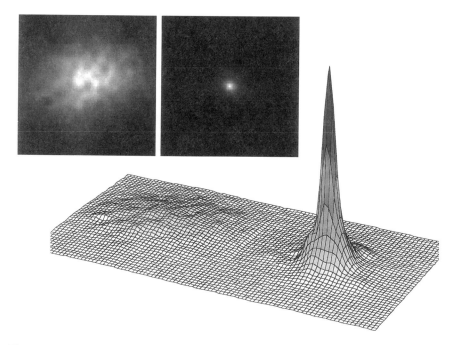

Figure 11.2 Comparison of star images taken with the *Starfire Optical Range* 1.5 meter telescope through uncompensated turbulence (left image) and with laser beacon adaptive optics (right image). Image scale is 2.0 arc seconds square, imaging wavelength is 0.88 ± 0.05 µm, exposure time is one second for each image, stellar magnitude is 2.7, full width half maximum image sizes are 1.65 and 0.18 arc seconds, and the compensated image peak intensity is about 25% of the theoretical maximum (Strehl ratio = 0.25). The compensated image intensity is 14 times greater than the intensity of the uncompensated image. These are raw images, displayed with no processing. (*Starfire Optical Range, Phillips Laboratory.* Photo supplied by the Technical Director, Robert Q. Fugate; also printed in *Optics & Photonics New*, June 1993, p. 18; © Optical Society of America.)

inexpensive telescope guidance systems, and with the entire system connected to the amateur's home computer, he is able, once again, to provide valuable contributions. And now the profession comes up with laser beacons!

11.2 FRESNEL DIFFRACTION VIA FOURIER TRANSFORM

The visualization of the results of the coma aberration required the Fourier transformation of a complex function. It can be shown that the Fresnel diffraction amplitude function for a simple real aperture can be obtained similarly by Fourier transformation of $\exp(i\pi/\lambda z)(x^2 + y^2)$ which must then also be multiplied by two additional complex exponential functions. (λ is the wavelength and z is the distance from the diffracting aperture; this product can be scaled for the computer.) The complex square of this result yields the Fresnel irradiance and as such, the two additional complex functions can be ignored if one is seeking only the irradiance portrayal. We have not examined Figure 6.1 to see if convolution could also be a route to Fresnel irradiance, nor have we examined Fresnel diffraction by apertures with specific transmission and/or phase functions.

Here is a procedure for a Fresnel irradiance portrait of a simple square aperture where $\lambda z = \frac{1}{2}$:

```
[x,y]=meshdom(-1:2/19:1,-1:2/19:1);
fres=(x.^2+y 2)*(1);
ap=zeros(90);
ap(37:56,37:56)=ones(20).*exp(i*2*pi*fres);
ft=fftshift(fft2(fftshift(ap)));
fresirr=ft.*conj(ft);
mesh(fresirr)                           See Figure 11.3.
```

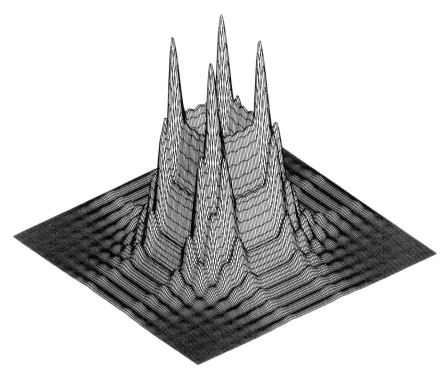

Figure 11.3 Fresnel irradiance from a square aperture, obtained by Fourier transformation, see text for scaling.

FRESNEL DIFFRACTION VIA FOURIER TRANSFORM   309

Is this not easier than using Cornu spirals and Fresnel integrals? Their use has recently been described by M. A. Heald (1986). If one wishes to have numerical data, it exits, here in the "fresirr" matrix, and likewise in all the other matrices created as described in this book.

The command mesh(ap) will show the real part of the fictitious aperture used for this calculation. The command mesh(ft) will show the real part of the transform which requires multiplication by two more complex functions in order to portray the Fresnel amplitude. As the quantity "fres" approaches zero, the irradiance pattern calculated above, "fresirr," approaches that of Fraunhofer diffraction.

In SURFER the data can be converted for a topographic plot (Figure 11.4) and it shows the "plaid" appearance often found in Fresnel patterns from rectangular structures. Contour line density is related to irradiance. Thus can Fresnel diffraction be portrayed for any simple aperture.

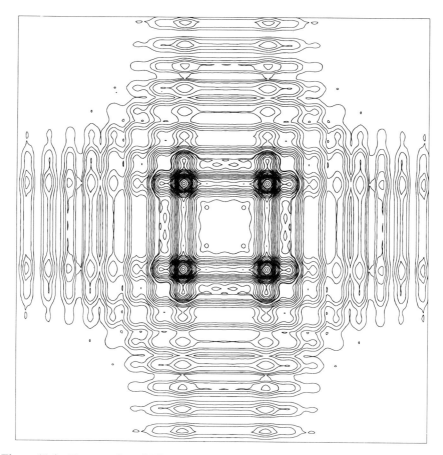

Figure 11.4 Topography of Figure 11.3; topographic levels not all equally spaced.

REFERENCES TO FRESNEL DIFFRACTION BY FOURIER TRANSFORMATION

G. O. Reynolds, J. B. DeVelis, G. B. Parrent, Jr., and B. J. Thompson, *The New Physical Optics Notebook: Tutorials in Fourier Optics*. SPIE—The International Society for Optical Engineering, and American Institute of Physics, Bellingham, WA, 1989, pp. 59–60.

J. W. Goodman, *Introduction to Fourier Optics*, McGraw-Hill, New York, 1968, p. 60.

J. W. Goodman, "Operations achievable with coherent optical information processing systems," *Proc. IEEE*, Vol. 65, pp. 29–38 (1977). Reprinted in: F. T. S. Yu and S. Yin, Eds., *Selected Papers on Coherent Optical Processing*, SPIE Optical Engineering Press, Bellingham, WA, 1992, pp. 270–279.

K. Iizuka, *Engineering Optics*, 2nd Edition, Springer-Verlag, New York and Berlin, 1987, pp. 62–68.

H. J. Weaver, *Applications of Discrete and Continuous Fourier Analysis*, Wiley-Interscience, New York, 1983, pp. 206–217.

REFERENCES

The mathematical level of entries marked thus, ■ is advanced.

G. Arfken, *Mathematical Methods for Physicists*, Academic, New York, 1970.
C. A. Balanis, *Antenna Theory: Analysis and Design*, Wiley, New York, 1982.
P. P. Banerjee and T.-C. Poon, *Principles of Applied Optics*, Irwin, Homewood, IL, 1991.■
K. R. Barnes, *The Optical Transfer Function*, Elsevier, New York, 1971.
R. H. T. Bates and M. J. McDonnell, *Image Restoration and Reconstruction*, Oxford, New York, 1986.
C. A. Bennett, "A computer-assisted experiment in single-slit diffraction and spatial filtering," *Am. J. Phys.*, Vol. 58, pp. 75–78 (1990).
J. M. Blackledge, *Quantitative Coherent Imaging—Theory, Methods and Some Applications*, Academic, London and San Diego, 1989.
M. Boas, *Mathematical Methods in the Physical Sciences*, Wiley, New York, 1983.
R. Bracewell, *The Fourier Transform and Its Applications*, McGraw-Hill, New York, 1965.■
H. J. J. Braddick, *Vibrations, Waves, and Diffraction*, McGraw-Hill, New York, 1965.
E. O. Brigham, *The Fast Fourier Transform*, Prentice-Hall, Englewood Cliffs, NJ, 1974.■
G. Brown, *Modern Optics*, Reinhold, New York, 1965.
J. C. Brown, "Fourier analysis and spatial filtering," *Am. J. Phys.*, Vol. 39, pp. 797–801 (1971).
R. Buckley, *Oscillations and Waves*, Student Monographs in Physics, Adam Hilger, Bristol, England, 1985.
F. P. Carlson, *Introduction to Applied Optics for Engineers*, Academic, New York, 1977.
M. Cartwright, *Fourier Methods for Mathematicians, Scientists and Engineers*, Horwood, Chichester, England, 1988.

D. C. Champeney, *Fourier Transforms and Their Physical Applications*, Academic, New York, 1973.

D. C. Champeney, *Fourier Transforms in Physics*, Student Monographs in Physics, Adam Hilger, Bristol, England, 1985.

D. C. Champeney, *A Handbook of Fourier Theorems*, Cambridge University Press, Cambridge, 1987.

X. Chen, J. Huang, and E. Loh, "Two-dimensional fast Fourier transform and pattern processing with IBM PC," *Am. J. Phys.*, Vol. 56, pp. 747–749 (1988).

V. Čížek, *Discrete Fourier Transforms and Their Applications*, Adam Hilger, Bristol, England, 1986.

C. A. Culver, *Musical Acoustics*, McGraw-Hill, New York, 1956.

R. W. Ditchburn, *Light*, Academic, New York, 1976.

S. A. Dodds, "An optical diffraction experiment for the advanced laboratory," *Am. J. Phys.*, Vol. 58, pp. 663–668 (1990).

P. M. Duffieux, *The Fourier Transform and Its Application to Optics*, Wiley-Interscience, New York, 1983.

E. Everhart and J. W. Kantorski, "Diffraction patterns produced by obstructions in reflecting telescopes of modest size," *Astron. J.*, Vol. 64, pp. 455–463 (1959).

D. S. Falk, D. R. Brill, and D. G. Stork, *Seeing the Light—Optics in Nature, Photography, Color, Vision, and Holography*, Wiley, New York, 1986.

G. B. Folland, *Fourier Analysis and Its Applications*, Wadsworth and Brooks/Cole, Pacific Grove, CA, 1992. ■

G. R. Fowles, *Introduction to Modern Optics*, 2nd Edition, Holt, Rinehart and Winston, New York, 1975. Reprinted by Dover, New York, 1989.

M. Françon, *Diffraction—Coherence in Optics*, Pergamon, Oxford, 1966.

M. Françon, *Optical Image Formation and Processing*, Academic, New York, 1979.

R. Q. Fugate, "Laser beacon adaptive optics," *Optics and Photonics News*, Vol. 4, June 1993, pp. 14–19.

J. D. Gaskill, *Linear Systems, Fourier Transforms, and Optics*, Wiley, New York, 1978. ■

E. V. Gillon Jr. (selector), *Geometric Design and Ornament*, Dover, New York, 1969.

R. C. Gonzalez and P. Wintz, *Digital Image Processing*, Addison-Wesley, Reading, MA, 1987.

J. W. Goodman, *Introduction to Fourier Optics*, McGraw-Hill, New York, 1968. ■

J. W. Goodman, "Operations achievable with coherent optical information processing systems," *Proc. IEEE*, Vol. 65, pp. 29–38 (1977). Reprinted in: F. T. S. Yu and S. Yin (Eds.), *Selected Papers on Coherent Optical Processing*, SPIE Optical Engineering Press, Bellingham, WA, 1992, pp. 270–279.

J. W. Goodman (Ed.), *International Trends in Optics*, Academic, San Diego, CA, 1991.

I. Grattan-Guinness, *Joseph Fourier 1768–1830*, The MIT Press, Cambridge, MA, 1972.

A. H. Greenway, "Proposal for phase recovery from a single intensity distribution," *Opt. Lett.*, Vol. 1, p. 10 (1977).

R. Guenther, *Modern Optics*, Wiley, New York, 1990. Chapters 6, 10.

A. Hairie and J. Provost, "Optical correlator using the incoherent light of a slide projector," *Am. J. Phys.*, Vol. 51, pp. 832–836 (1983).

G. Harburn, C. A. Taylor, and T. R. Welberry, *Atlas of Optical Transforms*, G. Bell & Sons Ltd, London, 1975.

R. Harding, *Fourier Series and Transforms—A Computer Illustrated Text*, Adam Hilger, Bristol, England, 1985.

A. P. Harrington and A. T. Winter, "Aperture plates for Fraunhofer diffraction," *Eur. J. Phys.*, Vol. 5, pp. 238–242 (1984).

M. A. Heald, "Computation of Fresnel diffraction," *Am. J. Phys.*, Vol. 54, pp. 980–983 (1986).

O. S. Heavens and R. W. Ditchburn, *Insight into Optics*, Wiley, New York, 1991.

E. Hecht, *Optics*, 2nd Edition, Addison-Wesley, Reading, MA, 1987. Chapters 11, 14.

J. Herivel, *Joseph Fourier—The Man and the Physicist*, Clarendon Press, Oxford, England, 1975.

H. H. Hopkins, *Wave Theory of Aberrations*, Oxford University Press, London, 1950.

S. Horemis, *Optical and Geometrical Patterns and Designs*, Dover, New York, 1970.

J. L. Horner and P. D. Gianino, "Phase-only matched filtering," *Appl. Opt.*, Vol. 23, pp. 812–816 (1984).

J. N. Howard (Ed.), *Optics Today*, Readings from "Physics Today," American Institute of Physics, New York, 1986.

K. Iizuka, *Engineering Optics*, 2nd Edition, Springer-Verlag, New York and Berlin, 1987.

P. Jacquinot and B. Roizen-Dossier, "Apodisation," *Progress in Optics*, Vol. III, North Holland, Amsterdam, 1964.

R. C. Jennison, *Fourier Transforms and Convolutions for the Experimentalist*, Pergamon, New York, 1961.

D. R. Kaiser and B. R. Russell, "Demonstrations of the phase reversal effect in Fraunhofer diffraction patterns," *Am. J. Phys.*, Vol. 48, pp. 674–675 (1980).

W. Kaplan, *Advanced Calculus*, Addison-Wesley, Reading, MA, 1952.

C.-J. Kim and R. R. Shannon, "Catalog of Zernike Polynomials," Chapter 4 in: R. Shannon and J. Wyant (Eds.), *Applied Optics and Optical Engineering*, Vol. X, Academic, San Diego, 1987.

M. V. Klein and T. E. Furtak, *Optics*, 2nd Edition, Wiley, New York, 1986. Chapter 6.

T. W. Körner, *Fourier Analysis*, Cambridge University Press, Cambridge, England, 1988.■ (But in between the advanced math it is a delight to read.)

J. D. Kraus, *Electromagnetics*, McGraw-Hill, New York, 1984.

E. Kreyszig, *Advanced Engineering Mathematics*, Wiley, New York, 1972.

J. Larcher, *Optical and Geometrical Allover Patterns*, Dover, New York, 1979.

E. N. Leith and J. Upatnieks, "Reconstructed wavefronts and communication theory," *J. Opt. Soc. Am.*, Vol. 52, pp. 1123 (1962).

L. Levi, *Applied Optics: A Guide to Optical System Design*, Vol. 2, Wiley, New York, 1980.

H. Lipson, *Optical Transforms*, Academic, New York, 1972.

S. G. Lipson and H. Lipson, *Optical Physics*, Cambridge University Press, London, 1969.

H. Lipson and C. A. Taylor, *Fourier Transforms and X-Ray Diffraction*, G. Bell, London, 1958.

R. K. Livesley, *Mathematical Methods for Engineers*, Ellis Horwood Ltd., Chichester, West Sussex, England (Halsted Press, New York), 1989.

R. S. Longhurst, *Geometrical and Physical Optics*, 2nd Edition, Longmans, London, 1967.

D. J. Lovell, *Optical Anecdotes*, Society of Photo-Optical Instrumentation Engineers (SPIE), Bellingham, WA, 1981.

V. N. Mahajan, *Aberration Theory Made Simple*, SPIE Optical Engineering Press, Bellingham, WA, 1991.

J. B. Marion, *Classical Electromagnetic Radiation*, Academic, New York, 1965.

D. McLachlan Jr., "The role of optics in applying correlation functions to pattern recognition," *J. Opt. Soc. Am.*, Vol. 52, pp. 454–459 (1962).

A. B. Meinel, M. P. Meinel, and N. J. Woolf, "Multiple aperture telescope diffraction images." Chapter 5 in: *Applied Optics and Optical Engineering*, Vol. IX, Academic Press, New York, 1983, pp. 149–201.

Y. Meyer, *Wavelets*, Cambridge University Press, New York, 1993.

J. P. Mills and B. J. Thompson, "Effect of aberrations and apodization on the performance of coherent optical systems. I. The amplitude impulse response," *J. Opt. Soc. Am. A*, Vol. 3, pp. 694–703 (1986).

K. D. Möller, *Optics*, University Science Books, Mill Valley, CA, 1988.

L. Mosley, *Hirohito—Emperor of Japan*, Prentice-Hall, Englewood Cliffs, NJ, 1966.

A. Nussbaum and R. A. Phillips, *Contemporary Optics for Scientists and Engineers*, Prentice-Hall, Englewood Cliffs, NJ, 1976. Chapter 10.

E. L. O'Neill, *Introduction to Statistical Optics*, Addison-Wesley, Reading, MA, 1963. ■

D. C. O'Shea, *Elements of Modern Optical Design*, Wiley-Interscience, New York, 1985.

H. Ōuchi, *Japanese Optical and Geometrical Art*, Dover, New York, 1977.

A. Papoulis, *Systems and Transforms with Applications in Optics*, McGraw-Hill, New York, 1968. ■

F. L. Pedrotti, S. J., and L. S. Pedrotti, *Introduction to Optics*, Prentice-Hall, Englewood Cliffs, NJ, 1987. Chapter 24.

M. J. Pérez-Ilzarbe, "Phase retrieval from the power spectrum of a periodic object," *J. Opt. Soc. Am. A*, Vol. 9, p. 2138 (1992).

Perkin-Elmer Corporation, "The practical application of modulation transfer functions," *Introductory Talks from a Perkin-Elmer Symposium*, March 6, 1963. (Copyright, Society of Photographic Scientists and Engineers, Inc., 1965).

L. J. Pinson, *Electro-Optics*, Wiley, New York, 1985.

M. C. Potter, *Mathematical Methods in the Physical Sciences*, Prentice-Hall, Englewood Cliffs, NJ, 1978.

R. W. Ramirez, *The FFT, Fundamentals and Concepts*, Prentice-Hall, Englewood Cliffs, NJ, 1985. © Tektronix.

G. O. Reynolds, J. B. DeVelis, G. B. Parrent Jr., and B. J. Thompson, *The New Physical Optics Notebook: Tutorials in Fourier Optics*, SPIE Optical Engineering Press, Bellingham, WA, 1989.

K. F. Riley, *Mathematical Methods for the Physical Sciences*, Cambridge University Press, London, 1974.

B. E. A. Saleh and M. C. Teich, *Fundamentals of Photonics*, Wiley Series in Pure and Applied Optics, Wiley-Interscience, New York, 1991.

R. V. Shack and J. E. Harvey, "An investigation of the distribution of radiation scattered by optical surfaces," *Optical Sciences Center (University of Arizona) Final Report*, August 1975. [See also "Harvey" in our Bibliography.]

A. R. Shulman, *Optical Data Processing*, Wiley, New York, 1970.

R. C. Smith and J. S. Marsh, "Diffraction patterns of simple apertures," *J. Opt. Soc. Am.*, Vol. 64, pp. 798–803 (1974).

F. G. Smith and J. H. Thompson, *Optics*, Wiley, New York, 1971.

M. R. Spiegel, *Mathematical Handbook of Formulas and Tables*, Schaum's Outline Series, McGraw-Hill, New York, 1968.

M. R. Spiegel, *Advanced Mathematics for Scientists and Engineers*, Schaum's Outline Series, McGraw-Hill, New York, 1971.

H. Stark, W. R. Bennett, and M. Arm, "Design considerations in power spectra measurements by diffraction of coherent light," *Appl. Opt.*, Vol. 8, pp. 2165–2172 (1969).

E. G. Steward, *Fourier Optics: An Introduction*, Horwood, Chichester, England, 1983.

W. L. Stutzman and G. A. Thiele, *Antenna Theory and Design*, Wiley, New York, 1981.

F. Su (Ed.), *Technology of Our Times: People and Innovation in Optics and Optoelectronics*, SPIE, Bellingham, WA, 1990.

C. Taylor, *Diffraction*, Student Monographs in Physics, Adam Hilger, Bristol, England, 1987.

C. A. Taylor, *Images—A Unified View of Diffraction and Image Formation with all Kinds of Radiation*, The Wykeham Science Series, Wykeham, London, 1978.

C. A. Taylor and H. Lipson, *Optical Transforms—Their Preparation and Application to X-Ray Diffraction Problems*, G. Bell, London, 1964. (Also Cornell University Press, 1965.)

H. F. A. Tschunko and P. J. Sheehan, "Aperture configuration and imaging performance," *Appl. Opt.*, Vol. 10, pp. 1432–1438 (1971).

R. K. Tyson, *Principles of Adaptive Optics*, Academic, San Diego, 1991.

A. VanderLugt, *Optical Signal Processing*, Wiley-Interscience, New York, 1992.

R. A. Waldron, *Waves and Oscillations*, Van Nostrand, Princeton, NJ, 1964.

J. G. Walker, "The phase retrieval problem—a solution based on zero location by exponential apodization," *Optica Acta*, Vol. 28, pp. 735–738 (1981).

J. S. Walker, *Fourier Analysis*, Oxford University Press, New York, 1988. ■

J. S. Walker, *Fast Fourier Transforms*, CRC Press, Boca Raton, 1991.

F. E. Washer and I. C. Gardner, *Method for Determining the Resolving Power of Photographic Lenses*, National Bureau of Standards Circular 533, U.S. Government Printing Office, May 20, 1953.

H. J. Weaver, *Applications of Discrete and Continuous Fourier Analysis*, Wiley-Interscience, New York, 1983.

R. H. Webb, *Elementary Wave Optics*, Academic, New York, 1969.

W. T. Welford, *Aberrations of Optical Systems*, Adam Hilger, Bristol, England, 1986.

W. T. Welford, *Optics*, Oxford University Press, Oxford, 1988.

C. S. Williams and O. A. Becklund, *Optics: A Short Course for Engineers & Scientists*, Wiley-Interscience, New York, 1972.

C. S. Williams and O. A. Becklund, *Introduction to the Optical Transfer Function*, Wiley-Interscience, New York, 1989.

R. G. Wilson, S. M. McCreary, and J. L. Thompson, "Optical transformations in three-space: simulations with a PC," *Am. J. Phys.*, Vol. 60, pp. 49–56 (1992).

C. W. Wong, *Introduction to Mathematical Physics—Methods and Concepts*, Oxford University Press, New York, 1991.

C. R. Wylie and L. C. Barrett, *Advanced Engineering Mathematics*, McGraw-Hill, New York, 1982.

F. T. S. Yu, *Introduction to Diffraction, Information Processing, and Holography*, MIT, Cambridge, 1973.

F. T. S. Yu and E. Y. Wang, "Undergraduate coherent optics laboratory," *Am. J. Phys.*, Vol. 41, pp. 1160–1169 (1973).

F. T. S. Yu and S. Yin (Eds.), *Selected Papers on Coherent Optical Processing*, SPIE Optical Engineering Press, Bellingham, WA, 1992.

COLLEGE LEVEL OPTICS TEXTBOOKS WHICH CONTAIN MATERIAL ON FOURIER METHODS

P. P. Banerjee and T.-C. Poon (1991); R. W. Ditchburn (1976); Fowles (1975); R. Guenther (1990); O. S. Heavens and R. W. Ditchburn (1991); E. Hecht (1987); K. Iizuka (1987); M. V. Klein and T. E. Furtak (1986); K. D. Möller (1988); A. Nussbaum and R. A. Phillips (1976); F. L. Pedrotti, S. J., and L. S. Pedrotti (1987); F. G. Smith and J. H. Thompson (1971); W. T. Welford (1988).

A SELECTED BIBLIOGRAPHY

Fred Abbott, "The evolution of the transfer function," Part I: concepts and definitions, *Optical Spectra*, March 1970, p. 54; Part II: the calculation of transfer functions, *Optical Spectra*, April 1970, p. 64.

Silverio P. Almeida and Hitoshi Fujii, "Fourier transform differences and averaged similarities in diatoms," *Appl. Opt.*, Vol. 18, p. 1663 (1979).

Jun Amako, Hirotsuna Miura, and Tomio Sonehara, "Wave-front control using liquid-crystal devices," *Appl. Opt.*, Vol. 32, p. 4323 (1993).

Pedro Andres, Carlos Ferreira, and Elvira Bonet, "Fraunhofer diffraction patterns from apertures illuminated with nonparallel light in nonsymmetrical Fourier transforms," *Appl. Opt.*, Vol. 24, p. 1549 (1985).

M. Arago, "Joseph Fourier - Biography read before the French Academy of Sciences by M. Arago," in the *Annual Report of the Board of Regents of the Smithsonian Institution, Showing the Operations, Expenditures, and Condition of the Institution for the Year 1871*, Government Printing Office, Washington, 1873.

Partha P. Banerjee, "A simple derivation of the Fresnel diffraction formula," *Proc. IEEE*, Vol. 73, p. 1859 (1985).

Partha P. Banerjee and Ting-Chung Poon, "On a simple derivation of the Fresnel diffraction formula and a transfer function approach to wave propagation," *Am. J. Phys.*, Vol. 58, p. 576 (1990).

Richard Barakat, "General diffraction theory of optical aberration tests, from the point of view of spatial filtering," *J. Opt. Soc. Am.*, Vol. 59, p. 1432 (1969).

R. Barakat, "The calculation of integrals encountered in optical diffraction theory," in: B. R. Frieden (Ed.), *The Computer in Optical Research*, Springer-Verlag, New York, 1980.

A. Barna, "Fraunhofer diffraction by semicircular apertures," *J. Opt. Soc. Am.*, Vol. 67, p. 122 (1977). [See the paper of Sethuraman and Srinivasan below.]

Stephen Berko, Yon G. Lee, Fulton Wright Jr., and Jon Rosenfeld, "Undergraduate laser diffraction experiments using large objects," *Am. J. Phys.*, Vol. 38, p. 348 (1970).

Ronald N. Bracewell, *The Hartley Transform*, Oxford Engineering Science Series Number 19, Oxford University Press, New York, 1986.

Ronald N. Bracewell, "Affine theorem for the Hartley transform of an image," *Proc. IEEE*, Vol. 82, p. 388 (1994).

Ronald N. Bracewell, "Aspects of the Hartley transform," *Proc. IEEE*, Vol. 82, p. 381 (1994). [See the paper of Villasenor below.]

R. N. Bracewell and J. D. Villasenor, "Fraunhofer diffraction by a spiral slit," *J. Opt. Soc. Am. A*, Vol. 7, p. 21 (1990).

H. J. Butterweck, "General theory of linear, coherent optical data-processing systems," *J. Opt. Soc. Am.*, Vol. 67, p. 60 (1977).

T. M. Connon and B. R. Hunt, "Image processing by computer," *Sci. Am.*, Vol. 245, October 1981, p. 214.

David Casasent, "Coherent optical pattern recognition," *Proc. IEEE*, Vol. 67, p. 813 (1979).

J. V. Cornacchio, "Transfer function of an elliptical annular aperture," *J. Opt. Soc. Am.*, Vol. 57, p. 1325 (1967).

Frank S. Crawford, "Comment on 'Fraunhofer diffraction patterns from apertures illuminated with nonparallel light'," *Am. J. Phys.*, Vol. 48, p. 313 (1980). [See the paper of Klingsporn below.]

Devon G. Crowe, Joseph Shamir, and Thomas W. Ryan, "Sidelobe reduction in optical signal processing," *Appl. Opt.*, Vol. 32, p. 179 (1993).

Jeffrey A. Davis, Mark A. Waring, Glenn W. Bach, Roger A. Lilly, and Don M. Cottrell, "Compact optical correlator design," *Appl. Opt.*, Vol. 28, p. 10 (1989).

C. de Izarra and O. Vallee, "On the use of linear CCD image sensors in optics experiments," *Am. J. Phys.*, Vol. 62, p. 357 (1994).

Fred M. Dickey and Donald J. Moore, "White light optical processor for edge enhancement and spectral filtering," *Appl. Opt.*, Vol. 18, p. 1679 (1979).

Okan K. Ersoy, "A comparative review of real and complex Fourier-related transforms," *Proc. IEEE*, Vol. 82, p. 429 (1994).

F. A. Fischbach and J. S. Bond, "Fraunhofer diffraction patterns of microparticles," *Am. J. Phys.*, Vol. 52, p. 519 (1984).

Hitoshi Fujii and Silverio P. Almeida, "Coherent spatial filtering with simulated input," *Appl. Opt.*, Vol. 18, p. 1659 (1979).

Wolfgang Gilliar, William S. Bickel, Gordon Videen, and David Hoar, "Light scattering from fibers: an extension of the single-silt diffraction experiment," *Am. J. Phys.*, Vol. 55, p. 555 (1987).

Gerald J. Grebowsky, "Fourier transform representation of an ideal lens in coherent optical systems," *NASA Technical Report, TR R-319*, January 1970.

Franklin S. Harris, "Light diffraction patterns," *Appl. Opt.*, Vol. 3, p. 909 (1964).

Franklin S. Harris Jr., Michael S. Tavenner, and Richard L. Mitchell, "Single-slit Fresnel diffraction patterns: comparison of experimental and theoretical results," *J. Opt. Soc. Am.*, Vol. 59, p. 293 (1969).

James E. Harvey and Roland V. Shack, "Aberrations of diffracted wave fields," *Appl. Opt.*, Vol. 17, p. 3003 (1978).

James E. Harvey, "Fourier treatment of near-field scalar diffraction theory," *Am. J. Phys.*, Vol. 47, p. 974 (1979).

James E. Harvey, "Fourier integral treatment yielding insight into the control of Gibb's phenomena," *Am. J. Phys.*, Vol. 49, p. 747 (1981).

Richard E. Haskell, "Fourier analysis using coherent light," *IEEE Trans. on Educ.*, Vol. E-14, No. 3, p. 110 (1971).

R. M. Herman, John Pardo, and T. A. Wiggins, "Diffraction and focusing of Gaussian beams," *Appl. Opt.*, Vol. 24, p. 1346 (1985).

R. J. Higgins, "Fast Fourier transform: an introduction with some minicomputer experiments," *Am. J. Phys.*, Vol. 44, p. 766 (1976).

Yoshiaki Horikawa, "Resolution of annular-pupil optical systems," *J. Opt. Soc. Am.* A, Vol. 11, p. 1985 (1994).

Mark Hose, "FFT chips for transform-based image processing," *Advanced Imaging*, June 1992, p. 56.

Guy Indebetouw and Chanin Varamit, "Spatial filtering with complementary source-pupil masks," *J. Opt. Soc. Am.* A, Vol. 2, p. 794 (1985).

D. Joyeux and S. Lowenthal, "Optical Fourier transform: what is the optimal setup?," *Appl. Opt.*, Vol. 21, p. 4368 (1982).

Suganda Jutamulia, "Phase-only Fourier transform of an optical transparency," *Appl. Opt.*, Vol. 33, p. 280 (1994).

Y. P. Kathuria, "Fresnel and far-field diffraction due to an elliptical aperture," *J. Opt. Soc. Am.* A, Vol. 2, p. 852 (1985).

Paul E. Klingsporn, "Fraunhofer diffraction patterns from apertures illuminated with nonparallel light," *Am. J. Phys.*, Vol. 47, p. 147 (1979).

Jerome Knopp and Michael F. Becker, "Virtual Fourier transform as an analytical tool in Fourier optics," *Appl. Opt.*, Vol. 17, p. 1669 (1978).

Jiří Komrska, "Simple derivation of formulas for Fraunhofer diffraction at polygonal apertures," *J. Opt. Soc. Am.*, Vol. 72, p. 1382 (1982). [See the paper of Saga and Tanaka below.]

Katarina Kranjc, "Simple demonstration experiments in the Abbé theory of image formation," *Am. J. Phys.*, Vol. 30, p. 342 (1962).

Hal G. Krauss, "Huygens–Fresnel–Kirchoff wave-front diffraction formulations for spherical waves and Gaussian laser beams: discussion and errata," *J. Opt. Soc. Am.* A, Vol. 9, p. 1132 (1992).

B. P. Lathi, *Linear Systems and Signals*, Berkeley-Cambridge Press, Carmichael, CA, 1992.

Adolf W. Lohmann and David Mendlovic, "Self-Fourier objects and other self-transform objects," *J. Opt. Soc. Am.* A, Vol. 9, p. 2009 (1992).

Adolf W. Lohmann and David Mendlovic, "Image formation of a self-Fourier object," *Appl. Opt.*, Vol. 33, p. 153 (1994).

A. I. Mahan, C. V. Bitterli, and S. M. Cannon, "Far-field diffraction patterns of single and multiple apertures bounded by arcs and radii of concentric circles," *J. Opt. Soc. Am.*, Vol. 54, p. 721 (1964).

Manuel Martínez-Corral, Pedro Andrés, and Jorge Ojeda-Castañeda, "On-axis diffractional behavior of two-dimensional pupils," *Appl. Opt.*, Vol. 33, p. 2223 (1994).

Heidi Jo Marvin, "Fraunhofer diffraction by diamond-shaped apertures: a theoretical and experimental study," *Am. J. Phys.*, Vol. 56, p. 551 (1988).

T. W. Mayes and B. F. Melton, "Fraunhofer diffraction of visible light by a narrow slit," *Am. J. Phys.*, Vol. 62, p. 397 (1994).

Bruce Mechtly and Albert A. Bartlett, "Graphical representations of Fraunhofer interference and diffraction," *Am. J. Phys.*, Vol. 62, p. 501 (1994).

R. P. Millane, "Analytic properties of the Hartley transform and their implications," *Proc. IEEE*, Vol. 82, p. 413 (1994).

R. L. Mitchell, "A computerized 3-D plotting program," *Proc. SPIE*, Vol. 10, p. 44 (1967).

Peter E. Mueller and George Reynolds, "Image restoration by removal of random-media degradations," *J. Opt. Soc. Am.*, Vol. 57, p. 1338 (1967).

Nobuharu Nakajima, "Reconstruction of phase objects from experimental far field intensities by exponential filtering," *Appl. Opt.*, Vol. 29, p. 3369 (1990).

M. Nazarathy and J. Shamir, "Fourier optics described by operator algebra," *J. Opt. Soc. Am.*, Vol. 70, p. 150 (1980).

Kraig J. Olejniczak and Gerald Thomas Heydt, "Scanning the special section on the Hartley transform," *Proc. IEEE*, Vol. 82, p. 372 (1994). [See the paper of Villasenor below.]

Edward L. O'Neill, "Transfer function for an annular aperture," *J. Opt. Soc. Am.*, Vol. 46, p. 285 (1956); and Errata: *J. Opt. Soc. Am.*, Vol. 46, p. 1096 (1956).

Alan V. Oppenheim, "The importance of phase in signals," *Proc. IEEE*, Vol. 69, p. 529 (1981).

B. J. Pernick, "An optimum lens configuration for optical spatial filtering," *Am. J. Phys.*, Vol. 39, p. 959 (1971).

Laxman G. Phadke and Jim Allen, "Diffraction patterns for the oblique incidence gratings," *Am. J. Phys.*, Vol. 55, p. 562 (1987).

Cyril A. Pipan, "Alignment of a spatial filtering system," *Optical Spectra*, November/December 1968, p. 50.

Srisuda Puang-ngern and Silverio P. Almeida, "Converging beam optical Fourier transforms," *Am. J. Phys.*, Vol. 53, p. 762 (1985).

George O. Reynolds and David J. Cronin, "Imaging with optical synthetic apertures (Mills-Cross analog)," *J. Opt. Soc. Am.*, Vol. 60, p. 634 (1970).

J. Elmer Rhodes Jr., "Analysis and synthesis of optical images," *Am. J. Phys.*, Vol. 21, p. 337 (1953).

P. Rochon, T. J. Racey, and M. Zeller, "Apodization effects in small angle Fraunhofer diffraction from a thin metallic edge," *Am. J. Phys.*, Vol. 56, p. 559 (1988).

Nobuhiro Saga and Kazumasa Tanaka, "Relationship between the Abbé transform and the Maggi–Rubinowicz transformation formulas," *J. Opt. Soc. Am. A*, Vol. 3, p. 1450 (1986).

A. B. Schultz, T. Vernon Frazier, and E. Kosso, "Sonine's Bessel identity applied to apodization," *Appl. Opt.*, Vol. 23, p. 1914 (1984).

J. Sethuraman and R. S. Sirohi, "Analogy between Fraunhofer diffraction of certain apertures and SSB modulation with extension to multiple apertures," *J. Opt. Soc. Am.*, Vol. 70, p. 146 (1980).

J. Sethuraman and V. Srinivasan, "Problem: Fraunhofer diffraction at a semicircular aperture," *Am. J. Phys.*, Vol. 50, pp. 609 and 656 (1982).

Roland V. Shack, "Characteristics of an image-forming system," *J. Res. National Bureau of Standards*, Vol. 56, No. 5, May 1956.

Roland V. Shack, "Outline of practical characteristics of an image-forming system," *J. Opt. Soc. Am.*, Vol. 46, p. 755 (1956).

Thomas. M. Sheahen, "Importance of proper phase analysis in using Fourier transforms," *Am. J. Phys.*, Vol. 44, p. 22 (1976).

C. J. R. Sheppard and M. Hrynevych, "Diffraction by a circular aperture: a generalization of Fresnel diffraction theory," *J. Opt. Soc. Am. A*, Vol. 9, p. 274 (1992).

W. Sillito, "Fraunhofer diffraction at straight-edged apertures," *J. Opt. Soc. Am.*, Vol. 69, p. 765 (1979).

W. H. Steel, "A demonstration Fourierscope," *Appl. Opt.*, Vol. 9, p. 1721 (1970).

David E. Stoltzmann, "Multi-aperture diffraction," *Appl. Opt.*, Vol. 15, p. 21 (1976).

G. W. Swenson, "Synthetic-aperture radio telescopes," *Ann. Rev. Astron. Astrophys.*, Vol. 7, p. 353 (1969).

H. S. Tan, "Diffraction by nonplanar slit and circular apertures," *J. Opt. Soc. Am.*, Vol. 59, p. 1429 (1969).

William J. Thompson, "Fourier series and the Gibbs phenomenon," *Am. J. Phys.*, Vol. 60, p. 425 (1992).

Hubert F. A. Tschunko, "Derivation of the point spread function," *Appl. Opt.*, Vol. 22, p. 1413 (1983).

J. N. Turner, D. F. Parsons, and C. L. Andrews, "Diffraction of electromagnetic waves by transparent edges," *J. Opt. Soc. Am.*, Vol. 64, p. 789 (1974).

A. Vander Lugt, "Signal detection by complex spatial filtering," *IEEE Trans. on Inf. Theory*, vol. IT-10, p. 139, April 1964.

Anthony Vander Lugt, "Coherent optical processing," *Proc. IEEE*, Vol. 62, p. 1300 (1974).

John D. Villasenor, "Optical Hartley transforms," *Proc. IEEE*, Vol. 82, p. 391 (1994).

Christopher C. Wackerman and Andrew E. Yagle, "Phase retrieval and estimation with use of real-plane zeros," *J. Opt. Soc. Am. A*, Vol. 11, p. 2016 (1994).

Jearl Walker, "The amateur scientist—simple optical experiments in which spatial filtering removes the 'noise' from pictures," *Sci. Am.*, November 1982, p. 194.

T. A. Wiggins, "Hole gratings for optics experiments," *Am. J. Phys.*, Vol. 53, p. 227 (1985).

W. Witz, "Fraunhofer diffraction pattern of an aperture," *J. Opt. Soc. Am.*, Vol. 65, p. 1077 (1975).

INDEX

Adaptive optics, 304
Angular terminology, 2, 12, 14
Aperture functions convolved:
  complex, *see* Optical transfer
    function: modulation transfer
    function for complex apertures;
    phase transfer function for complex
    apertures
  real:
    circle blocked by square, 205
    circular, 151, 154–157
    circular ring, 214
    cosine apodized, 215
    cosine super-resolver, 223
    cross, 297
    diagonal squares, 166
    double circles, 239
    50% ring aperture, 239–240
    four circles, 237–238
    four squares, 236–237
    half square, half transmitting, 195–199
    separated circle and square, 158–160, 199
    square, 148–153, 191–194
    triangle, 200–203
    two squares, rectangles, 162–165
    well-proportioned T, 299–300
Aperture functions Fourier transformed:
  complex:
    astigmatism, 268
    coma, 262
    focus error, 272

    half aperture π phase lag, 245
    half aperture $\pi/2$ phase lag, 247
    linear phase, 250
    V-phase, 254
    Zernicke polynomial number 34, 273
  real:
    annular circular, 139, 203
    circle blocked by square, 205–207
    circular, 177–181, 196, 212
    circular ring, 213
    cosine apodized, 214–218
    cosine super-resolver, 220–224
    cross, 296–299
    four circles, 237–238
    four squares, 233–237
    half square, half transmitting, 194–199
    linear taper apodized, 216–220
    parallelogram, 180–186
    rectangular, 116–117, 175–177
    square, 107–116, 172–175, 186–191
    triangle, 200–204
    two separated squares, 160–161, 231–232
    two-slit super-resolver, 224–225
    well-proportioned T, 299–302
Apertures, 106
  functions, 106
  source books for photographing, 302
  where to buy, 241, 303
Aperture stop, 141
Apodization, 211
  cosine, 214

324 INDEX

Apodization (*Continued*)
  linear taper, 216
Atlas of Optical Transforms, 290
Average value, 26

Bessel function, 177–178

Cochlea, 7, 9
Coherent light, 134
  imaging, 132, 145
  optical transfer function (*OTF*), 138, 147, 165
  spread function (amplitude impulse response), 132–133, 137, 145
Convolution, 148
Cut-off spatial frequency, 154

Diffraction, 99–107
Dirichlet conditions, 23

Euler identities, 25

Fourier, Count Jean Baptiste Joseph, biography and portrait, 14–20
Fourier series:
  coefficients, 26
  definition, 23, 25
  examples:
    for always positive shifted square wave, 74
    for always positive square wave, 55
    for sawtooth function, 5
    for square wave, 27
  linearity, 37
Fourier transform pair, 1-space, 87
  sine, cosine, exponential, 87–88
Fourier transform properties:
  preliminary list, 2-space, 127–129
  separation, 107
  theorems, 2-space, 278–280
Fourier transforms, 1-space:
  for shifted square pulse, 92–96
  for square pulse, 89–91
Fourier transforms, 2-space, *see* Aperture functions Fourier transformed: complex; real
Fraunhofer approximation, 103
Frequency:
  1-space:
    negative, 41
    spacing, 57, 65–74
    spatial, 10
    spatial angular, 12
    temporal, 2, 3

  temporal angular, 12
  2-space:
    spatial, 109
    spatial angular, 109
Fresnel diffraction, 105
  via Fourier transformation, 307
Functions, even and odd, 52

Gibb's phenomena, 30

Harmonics, 6, 8
Hirohito, from biography of, 282
Huygens wavelets, 100–101

Imaging:
  with coherent light, 132, 146
  with incoherent light, 133, 142
Impulse response:
  amplitude, 137, 161
  intensity, 137, 161
Incoherent light, 134
  imaging, 133, 142
  optical transfer function (*OTF*), 138, 143, 162
  spread function (intensity impulse response), 135–138, 145–146
Intensity, 125–126, 133–134
Inverse problem, 291
Irradiance, 111, 121, 125–126

Light, amplitude of light wave, 125–126
Lovell, D. J., his biography of Fourier, 16

MATLAB, address, 208
Melles Griot WaveAlyzer, 278
Modulation transfer function (*MTF*), *see* Optical transfer function, modulation transfer function
Multiple apertures, 229

Near-field diffraction, *see* Fresnel diffraction
Negative frequency, 41

Obliquity factor, 102
O-MATRIX, address, 208
Optical transfer function (*OTF*):
  coherent light, 138, 147, 165
  by convolution, 148, 162
  definition, 138–140, 143–144, 148–159, 191–192
  incoherent light, 138, 143, 162
  modulation transfer function (*MTF*):
    for complex apertures:
      astigmatism, 272

coma, 269
  focus error aberration, 274
  π/2 phase half aperture, 252
  V-linear phase, 260
  Zernicke polynomial number 34, 277
  definition, 191–192
phase transfer function (*PTF*):
  for complex apertures:
    astigmatism, 272
    coma, 270
    focus error aberration, 275
    π/2 phase half aperture, 252
    V-linear phase, 260
    Zernicke polynomial number 34, 277
  definition, 191–192
Optical transforms, *see also* Aperture functions Fourier transformed: complex and real
Atlas of, reference to and examples from, 290
Overtones, 6, 8

Period, 1-space:
  spatial, 10, 12
  temporal, 2, 6, 12
Phase, and phase shifts, 74–84
  in transforms, 92–96, 225
Phase in apertures, 243
Phase transfer function, *see* Optical transfer function, phase transfer function
Plotting, 3-space:
  with MATLAB, 186
  with SURFER, 172
Point spread function (*PSF*), 137
  coherent, 132–133, 137, 146
  incoherent, 135–138, 145–146
  incoherent imaging with, 145
  worked example, 161. *See also* Impulse response
PSIPLOT, address, 208
Pupils, entrance and exit, 141

Resolution, 151–159

Shift, function, 48–53
  application to multiple apertures, 229
  effect on transform, 225
Software for plotting:
  MATLAB, 186, 208
  O-MATRIX, 208
  PSIPLOT, 208
  SURFER, 172, 208
Spatial filtering:
  for detection of phase, 286
  edge detection by, 284
  examples from *Atlas of Optical Transforms*, 290
  laboratory set-up, 296
SPIE (The Society of Photo-Optical Instrumentation Engineers), 14
Super-resolution, 211
  cosine, 220
  double aperture, 224
SURFER, address, 208

Topographic plotting, 185, 194, 197, 202, 206, 218, 222, 259, 309

Wavelength, 6, 10
Wave, traveling, 2, 12–13